'A moving, intriguing and beautifully conceived exploration of place, person and planet through time, *Earthed* speaks to the struggles of holding on during dark days and the power of hope in hard times.'
**Rob Cowen, author of *Common Ground***

'I'm in awe of this extraordinary book. Artfully crafted yet full of raw honesty, *Earthed* is unlike anything I've read before.'
**Lucy Jones, author of *Losing Eden***

'A beautiful memoir of a scattered mind and how it can find peace in the soil. Rebecca Schiller's gaze is unflinching and full of truth. So many readers will find themselves in these pages.'
**Katherine May, author of *Wintering***

'A stirring, powerful and honest examination of mental health, motherhood and the pressures we put on one another. A vital book.'
**Alice Vincent, author of *Rootbound***

'A hard and beautiful read. The tough truth about the simple life.'
**Eva Wiseman**

'*Earthed* is Rebecca Schiller's powerful, poetic meditation on the process of falling apart, and her love letter to the land that rooted and rebuilt her. A deeply affecting read.'
**Leah Hazard, author of *Hard Pushed***

'So powerful, so human and so compelling. This is a beautifully written story of land and life and people. How we need them all yet all struggle with that need. I couldn't put it down. Connecting past to present and future, this frank and vulnerable memoir is filled with hope, strength and resilience. A must read.'
**Frances Tophill, *Gardeners' World***

'A lyrical journey through nature and the human heart.'
**Sarah Langford, author of *In Your Defence***

'*Earthed* broke me open. Painful, visceral and amazing writing that I loved from start to finish. This is a book that I will read again and again.'
**Grace Timothy, author of *Lost in Motherhood***

'An intimate story of fragility and losing control . . . I loved the evocation of women from history, their reminder that trees and terrain are part of what we all traverse and part of what we need to look after ourselves.'
**Jessica Moxham, author of *The Cracks That Let the Light In***

'We can never really know the inner landscapes of another person's mind. In Rebecca Schiller's *Earthed*, we see the compelling portrait of a woman who struggles and pivots, persists and adapts to a mental health diagnosis in honest and insightful prose.'
**Kathryn Aalto, author of *Writing Wild***

'Profoundly observed and beautifully rendered, this is a timely reflection on what it means to be human, and the redemptive power of nature. It is both exquisitely personal and painfully universal, I was underlining whole paragraphs as I read. A remarkable book.'
**Charlotte Philby, author of *A Double Life***

'*Earthed* is the most beautiful, poetic book about what it means to be a woman raising kids in this high-pressured modern world while attempting to connect with the natural world in order to feel more grounded.'
**Annie Ridout, author of *Shy***

'An extraordinary, life-changing read. Honest, vulnerable and deeply moving.'
**Sara Venn, founder of Incredible Edible, horticulturist and food activist**

'A brave and honest memoir.'
**Alice O'Keeffe, author of *On the Up***

'Invoking the power of the land and the women who worked and walked it before, *Earthed* is a spellbinding account of an urgent search for wholeness, acceptance and belonging.'
**Andrew O'Brien, 'Gardens, Weeds and Words'**

'*Earthed* beautifully captures the unravelling of a woman and mother in all its untidy, unyielding and brutal reality. The honesty and rawness in the way Rebecca writes about her rage spoke to me so much. A powerful and poetic look at our connection to country, to those women who came before us and the understanding our own minds.'
**Penny Wincer, author of *Tender***

'An incredible, candid memoir: full of flowers, truth and the reality of growth.'
**Cariad Lloyd, creator of the Griefcast podcast**

We hope you enjoy this book. Please return or renew it by the due date.

You can renew it at www.norfolk.gov.uk/libraries or by using our free library app.

Otherwise you can phone 0344 800 8020 - please have your library card and PIN ready.

You can sign up for email reminders too.

# Earthed

Rebecca Schiller

Elliott&Thompson

First published 2021 by
Elliott and Thompson Limited
2 John Street
London WC1N 2ES
www.eandtbooks.com

ISBN: 978-1-78396-549-6

Permissions:
Page 9: 'The Hummingbird' from *Blue Horses* by Mary Oliver, published
by The Penguin Press New York. Copyright © 2014 by Mary Oliver.
Reprinted by permission of The Charlotte Sheedy Literary Agency Inc.

Page 225: "missing you" by Tania Hershman, published in *Terms and
Conditions* (Nine Arches Press, 2017). www.ninearchespress.com

9 8 7 6 5 4 3 2 1

A catalogue record for this book is available from
the British Library.

Typesetting by Marie Doherty
Printed in the UK by CPI Group (UK) Ltd,
Croydon, CR0 4YY

*For Jared, with love and faith*
*This is my letter – you'll open it when it suits you best.*


*For Clare, with gratitude*
*You are pure carbon. Never doubt it.*

# CONTENTS

# FEBRUARY 2020

TEST START

Blue square, red square, red circle, blue circle, red square, red square: CLICK.

This feels like a race that I have lost from the start. My body is alive with potential movements but I try not to give in to them. I'm aware of everything all at once, which is too much and makes it hard to concentrate on the – blue square, red circle, red circle: CLICK. My thoughts bounce off the red circle and into the future when I tell Jared about the test: this result, that result, which result do I want it to be? Blue circle, blue circle: CLICK. I need to speed my brain up or make the shapes slow down – blue circle, red square: NO CLICK – but how? If I could just rewind ten minutes and listen to the instructions again, perhaps it would be easier? That's what I'll do. I'll go just the tiniest half-step back and try harder to pin it all down.

The grey, ping-pong-sized ball is attached to the middle of my forehead with a black sweatband. I'm wearing the child's size, and at first it holds my head in its grip comfortingly. As the minutes pass the band constricts until my temples bisect the circle of my skull with a line of future headache. I keep it on though: the adult size doesn't have a single-use cover and, according to the news, this sort of thing is becoming important. As we get ready for the test to start, adjusting the equipment squeezed close together in the corner of the consulting room, the psychiatrist's assistant and I dance around this awkward issue of hygiene, unsure whether we are being overly cautious or not cautious enough.

I am sitting where I have been told to sit: facing a camera, at a laminate desk, in front of a laptop. There is no window and the room is small and impersonal: an examining couch behind me, clinical waste bins and a sink to my left. It has not yet got to the stage where we think to wash our hands.

The slickness in the well of my right palm as I take the black plastic button-clicker makes me realise how nervous I am. The assistant's instructions are simple. All I have to do is stay in this chair for twenty minutes pushing the button with my thumb every time the screen in front of me shows a matching shape sequence. Red square, red square: CLICK. Blue circle, red circle: NO CLICK. Blue circle, blue circle: CLICK. Blue square, blue circle: NO CLICK. Together the camera and ping-pong ball will record how much I move around, while the computer program keeps tabs on my accuracy and reaction times.

During the test the psychiatrist's assistant will sit behind me, just out of sight, making notes.

I have done much harder things than sitting in a too-warm room clicking, or not clicking, a button. Yet I'm breathing fast and my lungs are stretched taut – the soon-to-split skin of an over-watered tomato after a drought. I no longer know how to be in this kind of electric-lit absence of a room. None of the versions of myself that I usually wear to camouflage this seem to fit. My fingers shake slightly with this disorientation: the ruffle of grass in a June breeze, the tiny bounce of a birch branch as a goldfinch lands. To cover my discomfort I make three stupid jokes, which fall like stones on to the rubberised floor.

An example test starts. I yank my mind back to the laptop and find red and blue shapes sliding past my eyes very, very quickly. I struggle to catch them before they disappear. They shimmer and blend with each other and I am instantly jumbled and on the back foot.

Oh.

An hour ago, as my husband Jared drove me here, I watched the fields and forests of Kent turn into shopping centres and wondered whether my subconscious could skew the test's result one way or the other. Now that the shapes are in front of me, refusing to cooperate with my brain, I am realising that, as a patient, I don't have that kind of power. But once a try-hard, always a try-hard, and so I lean forwards, deepening my crow's feet, and squint with concentration. I ready myself by pushing my hair out of my eyes and shaking my tilted head sightly as if hoping to tip the fog out of my ears. There's a blond hair caught on the purple sapphire of my engagement ring: mine. A mane of two years of neglect now: elbow-length, with split ends and better-than-salon lightened by the sun.

Why didn't I tie it back? Why am I making myself do this? Why didn't I just stay at home?

This year I decided not to make resolutions. Not to wake every morning and tie each day to them and not to spend the evening crying because, once again, the wind had blown those twelve hours loose. 2019 was difficult, but so was the year before. And the one before that – 2017 – when we moved to the plot and more than my edges started to fray. So I wanted to find a new way to open myself to the future without expectation and unfold the promise of 2020 slowly and carefully.

Yet in my cold new-year garden, I couldn't help letting one hope in. As I went out to feed the hens, I spotted something in the mud. I told myself that the grey-green shoots were only grass but the buds revealed themselves anyway: tiny, tightly closed, spear tips that loosened into white bells. The snowdrops opened with an invitation to believe that, whatever this year would bring, it would be better than the one before.

And it will be.

And it won't.

TEST START

Blue square, red square, red circle, blue circle, red square, red square: CLICK.

This feels like a race that I have lost from the start. My body is alive with potential movements but I try not to give in to them. I'm aware of everything all at once, which is too much and makes it hard to concentrate on the – blue square, red circle, red circle: CLICK. My thoughts bounce off the red circle and into the future when I tell Jared about the test: this result, that result, which do I want it to be? Blue circle, blue circle: CLICK. There is a clock on the metal-and-glass shelf to my left. Its hands read 10:46 and I already wonder what time the test started and how many minutes have gone by. Blue circle, red square, blue circle, blue circle: CLICK. I hear the noise of biro on paper over my right shoulder: the assistant is writing something down. Anxiety prickles up my arms and into my throat as I wonder what notes she could possibly be taking already: I haven't done anything yet. I am sitting so still, clicking this button (as I have been told) in total silence – though of course the silence exists only outside my head. Blue circle, blue circle: CLICK.

My legs are very restless but I try to keep them fixed in one position. Red square, red square: CLICK. My mind is a kaleidoscope of blurry thoughts and questions: the task in hand, its interpretation, trying to read the woman sitting invisibly behind me, keeping all these tabs open, active and interlinked. Underneath this first avalanche of thought are many bigger-picture wonderings: about my family, my life, our way of life and, as these things pop up, they obscure parts of the red circle, blue circle, red square: NO CLICK.

I hear the receptionist leading someone into the room next door. Babble, footsteps, door open, door close. A muffled discussion makes it through the plasterboard between us and a single reassurance is audible: 'No one is trying to trip you up.' The receptionist speaks in a voice that I remember from the two sentences we exchanged earlier. Has this reassurance made the woman on the other side of the wall feel more nervous or less?

Blue square, blue square? CLICK? CLICK.

As I depress the button with my thumb again I realise that – ridiculously, embarrassingly – I am going to cry and there is probably nothing I can do about it. With as little movement as possible (red square, blue circle, blue circle: CLICK) I put my left hand on my right arm and discreetly dig my fingernails into the freckles there. I feel naked: underneath-my-skin-level naked. As if my fat and bones and partially digested food are suddenly visible and the world can see all the bloody parts of my body: the ones I usually dress up in skin and shove out into the world as if they were a person.

I am confused and ashamed, none of my sleights of hand work here and now a tear has escaped and is making its way down my right cheek. I decide to risk wiping it, disguising the movement as a nose scratch and treating myself to uncrossing and recrossing my legs at the same time. I need to stop this crying before the assistant notices, so, just for a second, I pretend there is night sky above me instead of ceiling tiles; a sliver of new moon in the west replacing the ugly brown water stain.

But the squares and circles block out the stars and I am sinking with the effort of separating them: NO CLICK. I've stopped wiping tears, so there's a regular beat of tiny, almost non-sounds as they hit my skirt.

Red circle, red square, blue square, red circle, blue, red, blue, red, square, circle, squircle, circed, squed, blircle.

Bled. Rue.

CLICK?

NO CLICK?

I am screwing this up and there doesn't seem to be anything I can do about it. No performance I can roll out and no illusion of competence. There's no one to hide behind here, no last-minute miracle push I can pull out of the bag, no way to fake it, no brilliant distraction, no covering humour, no meticulous preparation, no costume, no series of reminders and lists, no lie or excuse, no way to cancel at the last minute, no opt-out, no get-out, no convincing apology, no way to go back in time.

I have finally been caught out.

I take deliberate, slow breaths and try to overcome this realisation – blue square, blue square: CLICK – and become one with the – red, blue, blue, red, red – computer. I tell myself that I exist only for this button and these shapes. That nothing else matters. Except it does. It all matters. It is all connected. It is all important. Square, square, square, circle, square, circle, square. My house is square and our field is made of two rectangles. I guess the pond is a wonky circle? And inside the trees' trunks there are hundreds of them: concentric circles, red circles. A red circle? CLICK.

I am trying to stay in this room but I want to go home to the place where all my hopes and hurts are held safely by a boundary of trees and hedges, two acres of dirt and a swatch of red sky in the morning, blue sky at lunchtime and red sky at night. Red: shepherd's warning, shepherd's delight. Red sky, red circle, blue circle, blue circle: CLICK? Red, blue, red, blue, rose, delphinium, dahlia, anemone, bluebell, tulip, forget-me-not.

But I have forgotten.

Time has stopped flowing and become a pool of superglue. I am lost in it: held fast by primary colours. Beyond this screen I know there is a world where red circles are ladybird poppies I sowed in autumn and blue squares the sharp-edged cornflowers. In that place it is shadows and whispers that compel me to stare at the curves and corners of red and blue as they sway gently in the wind. There I watch a dronefly settle to take a little nectar and wait for the understanding to land in me, as it always does, in the tiniest fragments: pollen caught in bristles and then flown away across the many miles.

The plot is out of my reach though. Red square? Red square? I don't remember what a square is but I am sure I never want to see one again. I can't think why I have to click this button but I'm certain that I must. I click. I don't click. I click. CLICK. CLICK. CLICK.

I disappear.

I have become earth now. Crumbled to a fine tilth and blown into the cracks of the keyboard: lost between the 'A' and the 'S' and the 'D'. I am so fine and so light that I could plant carrots in myself and they would grow straight and true in my insubstantial being. Nothing exists but the squares and circles that keep coming at me relentlessly, an endlessness of commas, here in the place where I am stuck:

red, blue, red, blue, , , ,

The world as I know it has ended. The plot has died off, died back.

A felled forest: white noise and then silence.

And even the swirl of the wind is stopped.

## The Hummingbirds

In this book
   there are many hummingbirds—
the blue-throated, the bumblebee, the calliope
     the cinnamon, the lucifer, and of course
       the ruby-throated.

Imagine!
Well, that's all you can do.
For they're swift as the wind

and they fly, not across the pages but,
like many shy and otherworldly things,
   between them.
I know you'll keep looking now that I've told you.
I'm hungry to see them too, but I can't
   hold them back even for a moment, they're
     busy, as all things are, with their own lives.
So all I can do is let you know
   they're here somewhere.
All I can do is tell you
   by putting my own hunger on the page.

Mary Oliver, from *Blue Horses*

# PART ONE

# SPRING 2019

# I FALL

It is dark on this moonless night, the ground uneven, and the half-bottle of wine makes me forget to place each foot deliberately as I have trained myself to do. I am upright and laughing one moment and the next, side-lit by the pub window, I'm knees down on the wet pavement trying not to cry. Down here I spot mud and grass smeared on the side of my good shoes – a gift from my field no doubt. My palms are skinned and there's a pain somewhere else too, but I pay more attention to the two men in front of me turning their heads in interest and amusement.

'I'm fine. So like me. What a klutz,' I say to the friend I'm out with for the night, rolling my eyes faux-affectionately at myself and performing my best version of a capable, calm and unhurt woman with an unfortunate but cute habit of tipping herself on to the floor. It's a routine I do well. 'I'm okay, really, don't worry. Let's go on somewhere, have another drink,' I say to her, smiling, as I jack-in-a-box up. My friend has had a tough time of late and I know she needs me for an uncomplicated night out. As I walk through the narrow streets of the old town with her, I let myself feel the pain in my knee for a moment, and I can tell that I'm probably not so okay after all.

This fall is a thing to be carefully managed by my internal PR agency. I play a tactical game of deflection, telling my friend lightly about the four-year-old me who went to school and swiftly got given the 'girl with two left feet' tag. I paint a funny picture of my scruffy little self emerging from the prefab classroom dragging a manky teddy behind me; sweetly stuccoed with plasters, ripped clothes and undone plaits. I compress the story into three minutes, ending with four weeks ago when, attempting to show my children that I could leap, gazelle-like, from one wooden stepping stone to the other, like an idiot I fell in the adventure playground and bruised both knees so badly that I couldn't drive for a week. I draw a line under all this with minimising laughter and then return to celebrating her brave new life, which, with as much of my heart as I can muster, is what I am here for.

In the background, as I pay attention to her, I also think of the part of me that's still constantly bruising my hips and shins on the edges of furniture, ripping waistbands off jeans and destroying my dungarees by catching the straps on door handles. This me is still a four-year-old, unsure where the edges of her body are, where the floor and walls begin but not having any idea why.

I can't bear that I fell in front of strangers and an image of the men swivelling in amusement keeps popping up in my thoughts as we walk. A little vignette of embarrassment and self-loathing that will stay on a loop for much longer than it should, taking turns with the playback of my stepping-stone fall. That time no one noticed me hit the ground. Not the children, racing joyfully away from me, the backs of their fleeces covered in leaf mould, nor my husband sitting on the nearby bench looking at his muddy boots. I stayed down there for a while, my nose against the bark chippings for long enough to spot a woodlouse

creeping between two logs and to realise how like a tiny arma-
dillo it looked. As it worried its way into the woody darkness,
my shock and embarrassment coalesced into anger but, as ever,
I wasn't clear who it was I was angry with.

Later tonight, in a wine-bar bathroom, I sit on the cold
toilet seat and try to pull down my tights. They are stuck fast
to my skin: glued with blood from the four-inch gash that runs
from the bottom of my knee on a diagonal that splits my shin.
I instantly hate this cut and the little limp it is making me do.

So I cover it up. I go back out into the bar, glad of my long
skirt; I refuse to limp and am extra witty, extra clever and extra
outgoing. Later, lying on my friend's spare bed with the ceiling
spinning gently above me, I will think that while most people
are made of 60 per cent water, I am composed of smoke and
mirrors largely held together with shame.

The next morning I scrape the ice from my car before driving
back across the flat, reclaimed land that lies between my rare
night away and a two-acre patch of frozen mud, the smallholding
that Jared, our two children and I moved to in 2017. The light
is low and clouds hang heavily over the reeds that border the
water along the roadside. It's stark and beautiful with occasional
glimpses of swan and a liberal sprinkling of Romney sheep. In
summer we'll all walk lazily along here in the most English of
sunshine: spotting wildflowers, butterflies and bees and arguing
about whether 10 a.m. is too early for our picnic lunch.

Today, as I return to our plot, I notice the mess first, as
always: the weeds and grass that encroach on the gravel driveway,

piles of leaves we have found time to rake but not collect, a heap of broken bricks we plan to use as hardcore when we finally get round to filling in the flood-waiting-to-happen car-inspection pit dug by the previous owners. The paddock gate reveals a motley collection of field shelters, sheds and henhouses that contain an even motlier crew of geese, goats, chickens and ducks. Their comforting cacophony starts as I exit the car – desperate to let me know that they haven't yet been fed.

Even though I'm in my good clothes and nice shoes I walk straight through my garden towards the animal sheds. My borders offer the first glimpse of primroses telling winter to move along now. Hellebores and the spidery leaves of Anemone coronaria give welcome respite from the endless brown and my boots inadvertently trample a host of particularly keen daffodil shoots that ring the fruit trees in our little orchard. The season's early adopters are out in force this morning, but still it takes the faith I've been cultivating to believe in spring's coming in the sharp cold of this January day.

I pass the almost empty vegetable garden and tuck behind our beech hedge to let the animals into the field. The gang cheer me as they always do: a chuckle at the way the ducks chase our cats – beaks down, sheer gumption making them the unlikely winners in every encounter; the goats' distinct preferences for being stroked and scratched (Amber on her ear, Belle on the sides of her golden face) and their daily attempts to escape and nibble my rose cuttings – the hairy little gits. The white hen who streaks straight to the end of the field every morning to get into the hollow tree where she insists on laying.

Limping a little, I crouch down by the chicken shed to check if anyone remembered to collect the eggs last night and, as my knee bends, the new scab breaks, letting blood through. It barely hurts but I wince anyway with the thought of the night before. Out in today's morning quiet, I put my ungloved hands on the ground and try to be *here now* instead of *there then*. The earth feels cold: grass crispy with a light coating of frost, a present-moment shimmer camouflaging something more permanent below.

I brush a few of the frost crystals aside with my fingers and put more pressure on the field's surface. Today the frozen crust gives way easily, revealing only the wet clagginess beneath. My ground. My mud. It is why I am here: this clay, this land. To return to it, get back to it, to raise my children in it, to try naively to live the good life, the simple life, a life where this ground supports us.

Today, and every day this week, our plot is hard with the cold. The pond is a solid brown lump of ice, much to the surprise of the ducks, who waddle over to it every hour or so on a *Groundhog Day* loop of bewilderment. I find myself humming 'In the Bleak Midwinter' as I do my morning rounds with the animals and discover that the hose, water butt and all the buckets and drinkers have turned to solid ice. I replace the frozen drinkers with makeshift troughs and bowls (which will themselves freeze over in an hour) and take the others inside to start a constant cycle of thawing that leaves chicken-poo-infused puddles on the living-room floor.

As I finish my morning jobs I spot a blackbird with a distinctive necklace of featherless skin around his throat. There is something about the framing of this view, and the very same blackbird in it, that opens up a memory. The air is the same as it was then too: two years back on the day we moved here. The kind of deep cold that refuses to let the sun bring its usual warmth.

We were stuffed into the car: two adults, two children and two cats. Sparks of stress and excitement pinged off the sunroof, the windows and the box in the boot containing a kettle, cups and a corkscrew. We'd done this house-moving thing before and knew there would be an urgent need for hot caffeine on arrival, followed later by wine. The stench hit us about ten minutes into the journey. I'd daydreamed about this trip from our former home in a seaside town to the new house, just over an hour away on the edge of the Weald of Kent. I'd imagined the salty seaside air slowly mingling with countryside smells before disappearing entirely, every inhalation filling our noses with the iron tang of earth or perhaps an unseasonal breath of new-mown hay. An olfactory metaphor for the big adventure ahead. But before we'd even made it to the motorway, the smell in the car made me want to be sick.

We stopped in a lay-by, sniffed three-year-old Arthur warily and then opened the boot. The horror was definitely in there somewhere. We found him – the culprit – in his blue carrier, miaowing desperately. Perhaps we'd scooped him up and stuffed him into the car on his way to the litter tray. Maybe the stress of moving day had worked a kind of unholy magic on his feline bowels, but for whatever reason Bruce had smeared himself and the carrier with pungent catshit. It was all we could do not to leave him by the side of the road. Instead we cracked the windows, turned the heating up and I carried on driving, feeling a little chastened.

Then we were here, turning the car into the driveway. Jared got out before any of us, then the children and finally me. 'I can't believe that we live here, that this is all ours!' I said, though the words, once out in the freezing air, didn't sound quite right. But I moved on quickly, taking it all in: the plot, the work, the plans, the ideas already piling up and spreading out in my head. We walked along the path, past the log store, under the wooden arch that would soon fall down, and took out unfamiliar keys to open the porch for the first time. The stress of mortgages, solicitors, of packing and the journey dulled instantly. There's nothing like watching your children run in circles, shrieking with excitement and arguing about who gets which bedroom, to make you forget that ten minutes ago you were shouting and that your near future involves washing an angry cat. As we walked through the door now, this place that we had spent only minutes in before became our shelter and, though the kitchen was falling apart and the central heating didn't exist, every corner was transformed by the hopes we were piling in them. We were home.

A little later, despite chattering teeth, the children and I did our tenth excited loop around the plot, trying and failing to stay out of the way of Jared and the movers. We'd already clocked potential sites for chicken coops and dens by the time I spotted a row of leeks waiting to be harvested in the neat, otherwise bare, vegetable patch that sat (and still sits, somewhat less neatly) directly behind the south-facing house. A few hours later, our possessions stacked in each room and the removal lorry finally gone, I knew exactly what I wanted to do to mark the beginning of our lives in this squat bungalow with its wrap-around garden and tree-lined field – and it involved leeks.

I strolled over to the vegetable patch, my first real encounter with its soil squares, and felt consumed by the romance of harvesting produce sown nearly a year ago by the garden's previous caretaker. She had lived here for over thirty years and raised her family in this place where I would now raise mine. Our first meal in our new life would be a ritual dinner, a gift from the past. So I took hold of a leek by its peeling casing and pulled gently, savouring the poignant moment. Nothing happened. I tried again. This time I snapped off an inch or so of the inedible top, but was no closer to actually harvesting a leek.

Undeterred, I tried a new technique: grabbing it at the bottom where its thick white stem met the soil. I yanked really hard this time, but it didn't budge a millimetre. Was it a trick leek? I was flailing about by now, using every ounce of strength to get the damn thing out of the ground and feeling increasingly ridiculous. It would not yield to me. Fuck that fucking leek.

Twenty minutes later, having worked out where the movers had left the gardening tools, finding only three impossibly heavy spades – and something pointy and unfathomable – I improvised with a dessert spoon and set about digging. The soil was much harder than I'd expected so I was chipping away at it like just-out-of-the-freezer ice cream. Every now and then I gave the damn thing an experimental wiggle and shouted to my increasingly impatient family that everything was great, I wouldn't be long. What felt like hours later, I could finally see its gnarly base and the first signs of white worm-like roots clinging fast to the clods of clay. This was it; I would conquer this leek. Ha!

Red in the face and sweating profusely despite the January cold, I grabbed it once again and it seemed to give a little. I wiggled it and – yes – it yielded some more. I took it in both hands and, grunting with effort, pulled upwards from the depths of my being, from my soul. Finally, I felt the exquisite freedom of release as it gave up and came to me. And then snap. I was thrown backwards on to the hard ground. Only the tasteless green part was clutched in my red-raw hands and the white prize of its leeky, savoury goodness was still firmly one with the clay. Double fuck that fucking leek and the realisation that this was all going to be much harder than I'd thought.

At lunchtime today I am wearing my third-best dungarees over leggings, thick gloves and a ridiculous bobble hat. The cold air is a slap in the face as I go outside, but it distracts from the pain in my leg as I start digging close to where this year's leeks wait for a more successful harvest. That's for another week though, as right now there's a huge pile of woodchip that needs loading on to wheelbarrows, dumping on my paths and spreading evenly with a rake. It's a satisfying thing to cut through; parting easily enough with a whooshing sound of man-made meeting nature, and then a *shhhunk* as I lift the mulch and tip it into the barrow bed. One spade, two spade, three spade, four. It takes about twenty spadefuls to fill it and then, reckless, I usually gamble on another couple on the top. Back and forth, back and forth, I pass bare branches and juicy, duplicitous holly berries as I take out the crappy week I'm having on this mound of bark. I struggle not to break into a jog but I am trying to remember to walk rather than run between tasks. This overwhelming urge to keep moving so as to travel almost instantly from job to job, radically limiting any gaps in productivity, has been with me for as long as I can remember. No task is too big – it's just a matter of perspective and determination. I didn't learn to walk as a baby; I learned to sprint – swift but off balance.

Soon after we moved here – around the time that the full extent of the work ahead on this smallholding dawned on me – I tripled my pace rather than dial down on ambition or schedule. Within weeks my default speed became 'pretend I am being chased by a rabid dog and then go a bit faster'. And it did get things done, for a while at least. But there turned out to be consequences to keeping up a hysterical pace for too long: an abrupt stop. So now, even though my every instinct is to rush, I try to grit my teeth and take my time, telling myself that easing up means I can keep going for longer.

Finally, I finish covering this patch with a fresh layer of chippings and feel better. Daily pressures and nagging voices have receded. This task – the smoothing-over and hiding of imperfections – appeals to me. The quick pay-off, the miniature *Grand Designs* reveal, suits my natural impatience. I feel bad, I come outside, I see the sky, I dig a hole, I stroke a hen and I feel calmer. Every single time. This was a good idea.

Then I look to my right and see days-I-don't-haves'-worth of paths that also needing weeding and chipping. They actually look worse now that they have a tidy neighbour. And then there's the literal tonne of compost I've got to finish covering my veg and flower beds with. Not to mention the leaves smothering the crocuses and the grass growing in between every paving stone. Now the run is right there, ready to spring my ligaments, bones and muscles into action like a pent-up shout. I think about the seed trays I need to fill with compost to allow seedlings to push their way to the surface. Seeds still to buy. Poultry houses to disinfect. Worlds yet to conquer. Then thoughts of my children, Jared, the house, my work, the political mess, the world's slide towards climate disaster and everything else start jumping up and down and waving at me.

My hips, thighs and knees start to move. My breathing quickens.

I am running again.

Forty-eight miles away from me a civil servant is sitting at her desk worrying discreetly at a small bead bracelet held between finger and thumb. She is nervous, excited and the rhythm of her breath is more quavers than crotchets, though she tries to hide it, to calm herself – to fit in. She reads the pages on her screen with a face arranged into the perfect show of interested concentration. This is her first day in a new job and her first task – one that will take some months – is to produce the government's sixty-seventh annual report into smallholding. The sensory assault of the morning in a new office has left her tired and muddled. Head full of names that don't yet connect to faces, new rules – spoken and unspoken – the whirl of the journey, the need to be early, the moulding of her contours to fit the shape her predecessor left. She pauses to yawn behind her hand; thankful that the sound is disappeared by the judder-click of a printer and the slam of a taxi horn on the Westminster street below.

So far she has been reading last year's report slowly and carefully, preparing notes and questions to discuss with the team. But now she just lets the words about these council-owned parcels of land – 75, 150, 200, 300, 400 acres – wash around her brain. She thinks back to the weekend when, to celebrate this new job, she bought herself a copy of *The Smallholder* magazine. Over coffee, toast and the new silence of her Sundays, she looked at pictures of tinpot, piglet-strewn junkyards and genteel duck-egg-blue-painted cottages strung with bunting and decorated with jugs of home-grown flowers. She felt the pull of that life, as she often had before, and a crackle of excitement to be getting closer to it – at least in working hours. But the report she's reading now, and the one she'll soon have to write, don't quite seem to fit with the world she found in the magazine, nor the one always turning over in her imagination.

These slick set-ups are the real deal, of course, and they are more professional than what the clunky word – 'smallholding' – has always conjured for her. Even the mud seems shinier when six zeros of local-authority profit are lined up in the next column. Good, the civil servant tells herself. It is good to find something significant, when you were expecting three acres and an elderly cow. The case study she reads next is impressive. Four hundred and forty acres of wild flowers grown for seeds. Over the hum of the road she can almost hear the buzzing of the insects enjoying these vast new meadows. What a thing to be part of: this report that captures far more than her own hazy fantasy of hearing nothing in the early morning but soft breaths, starlings and rain on the roof.

She should feel happy and excited by this day-one change in perspective, and the bits of her not already dulled by strip lights and induction manuals do register the shift. Yet she's rolling the beads faster and faster, squeezing them harder against her fingerprints. And when she makes herself put the bracelet in her pocket, she pulls her scarf comfortingly around her shoulders like a shawl. Her hands drop to her belly; fingers fanning out over the rollercoaster loops of her fallopian tubes, the thin end of the wedge of her uterus and the slack of her cervix. She knows what it's like when something is missing and can't shake the feeling that the pixels in front of her now are incomplete. That the scale in this report is wrong. She's kidding herself with this reverence for largeness. It's the small things sometimes (often); the tiniest things – their presence, their absence – that have a world-smashing, big-bang-ing, universal impact. Boom!

Four hundred and forty acres. A union. A once-divided cell. A hope. The smallest of smallholdings.

Smallholding

Small. Holding.
This thing I have that is not much
and all too much for me.

'Small':
a pale yellow flower of a word on the stalk of the Old English
    'smæl'.
A thin meaning:
slender,
fine.

Under the soil a tangled root ball:
the proto-Germanic 'smal', the Gothic 'smalista', old Norse
    'smali',
church Slavonic 'malu'.

The original meaning of 'narrow' almost lost to us unless
we turn our tongues to waists or (worse)
intestines.

Sm–all.
Five letters, two sounds – all in, in 'small'.
Thirteenth-century diagraphs that pushed
out of an embryonic base to become:
not big, not large but
compact.

With decades' padding this little thing
got diminished:
a small person of small means and small import.
Small change, small talk, small fry,
small potatoes
– but when I dig them up, surprisingly heavy in my hand.

'Holding':
from the Middle English 'holden' to the Old English 'halden'.
Meanings peel away like layers
of the onions I pull from the earth:
sharper than you'd think with
a side-effect of tears.

To hold: containing something that would otherwise
escape: water, petrol, grain.
A gift, a load, a ritual
gathering.

A firm hold, on my arm:
ruling, controlling, imprisoning.

Or (on the other hand)
a cherishing touch
that watches while I rest.

Seven letters of: finding,
tending, keeping, owning.
Hard and soft sounds
that won't be put down.

*Hold on! Hold me!*
And [in a whisper]:
*never let me go.*
*(Let me go!)*

Last of all —
'a hold':
the void in a ship's
belly where the cargo is stored.

Empty and echoing.
Stuffed full of
flowers,

treasures,
breaths.

A little breath.
A small holding.
Holding.
Beholden.
Held.

In the short-changed light of the January afternoon – after finally washing my cut leg in the bath – I pull on a fresh set of work clothes and watch the birds going about their end-of-the-day business from my bedroom window. A dunnock (I think) on the hawthorn; blue tits having relay races from guttering to rose branch; a collared dove getting a head start on nest-building and my unimaginative favourite, the robin, hopping across the grass. I know this chap from all the others. He's charmingly pot-bellied with a tilt of his head that makes me think of a 1920s dancer with a top hat and cane: 'Robbo the Robin'. 'He's so fat, Mummy!' says Sofya, who's come into the room in her school uniform to hunt me down. Good old Robbo defuses my irritation at the feeling she's pursuing me and the swift guilt that follows. I don't want to be this person who reacts to my daughter as if she were an intruder in a rare five minutes of rest.

I kiss her on the top of her dark, tangled head and take a moment to breathe her in. Over the past couple of years my capacity for almost everything in life has shrunk dramatically. The generous, patient and unboundaried energy required to meet my children in the loud, needy, giving-of-myself places they often ask me to be is just not there any more. I cheat by taking them in like this: smelling them and letting that connect with the ancient bit of my brain that isn't so depleted, but which is very hard to find. I spend time stroking their cheeks, kissing their foreheads and saying loving things to them as they sleep, knowing that here in the dark they won't argue, talk incessantly or drop a fork on the floor – the clatter plucking every one of my nerves. I find a Canadian soap opera about a horse whisperer for us to watch together in almost silence, snuggled close, and I make it into an exclusive mother–child club. I give them what I can and hope it makes up for how little I have left.

Sofya and I also find a way to be together in our roles as Director and Deputy Director of Animal Husbandry. And now we are rushing to put on our boots, hats, gloves and coats and head out into the dimness that has abruptly claimed the afternoon. She mucks out the goats using a red dustpan and brush. I refill their hayrack and put hard feed in buckets; sprinkling it with supplement and listening to my daughter spill the contents of her mind into the little animal shelter we inherited. Always wanting to want to hear more of her sideways take on the world; always longing for a stretch of quiet.

As we walk back to the house I am already deep in the week ahead, but Sofya's voice breaks through: 'Look, Mummy!' I turn and find her behind me, absorbed by the sky. I was rushing again and nearly missed this January sunset laid out above the trees. But my eight-year-old moves at a slower pace; eyes open, ready to receive these kinds of gifts and this feels like a small victory. I walk back towards her smiling and put my arm around her shoulders. We stare at the red on the horizon and see it fade up to orange and then yellow. There's a band of green above, spreading like watercolour into the blue. Higher up the clouds are shocking pink and – behind witchy, black branches – the dusk is crushed berries. We say nothing – there is nothing that needs to be said as the animals' jaws work at their dinners, the birds sing their goodnights and our quiet breathing slowly synchronises.

The last day of the month and I walk across the garden's snowy ground as night gives into morning. Despite the smoothing of white over them, the unkempt hedges, flower beds and lawns are enormous today through eyes sharpened by overwhelm. This smallholding is, in reality, on the smallest of sides at a touch under two acres, but it has a way of expanding and contracting depending on how I am feeling and who is looking at it with me. The space we inhabit is vast through the eyes of our city friends yet poky when I come up with a new plan that won't fit. On my very worst days it is a universe whose edges I can never hope to touch – let alone weed. When our farmer neighbour Victor pushes a barrow between our yards, bringing over bales of hay, it all shrinks again to fit neatly in his pocket. A toy holding, not even a hectare and ineligible for the basic payments that even the tiniest of real farms receive from the government. With him I don't use the word 'smallholding'. I'm embarrassed to have called the place where we live and work anything but a garden with a few pets.

Now I let them out – my livestock, my pets – and watch them from the gate next to what, in summer, will be a grassy hollow but in winter months is an accidental pond. Today, this dip is hidden by snow and Honk the goose is walking slowly across it, hoping for a drink. As the water has turned to ice in the past fourteen hours her orange feet slap down comedically on a newly hard surface.

I've woken feeling tired and flimsy and need something more than the bite in the air to enliven me. A connection to something – a goose will do – might help me shake off this creeping feeling that something inside me has started to come loose. A cobblestone, agitated by generations of boot soles, working its way up to become a trip hazard.

There is nothing uneven about Honk though as she stretches; long and balanced, opening her wings in parallel to the ground, standing straight on her left foot and stretching her right out behind her: a perfect arabesque. I force myself to leave my head for a moment, eased down into my body by the view of her elegant form. I am an infrequent visitor here and it is uncomfortable to concentrate on the reality under my skin. Tight in the chest, a little dehydrated, achy in the lower back and the sticky constriction of a wound dressing on my leg.

I ended up at the GP surgery within a week of my pavement dive outside the pub. As the doctor peered professionally at the gross pus-y mess of my shin I was embarrassed by what, under his gaze, looked like lack of self-care: black hairs poking out of the weeping yellow and – how had I missed it? – a line of mud along my ankle bone.

There is not much I can do about the mud – a little of this earth is always on me now: a smear of boot edge catching the skin over my fibula, a grain of brown down the side of my nail that no amount of brushing can reach. I take the plot with me on the train to smart London meetings, on the school run, to my desk. There's dirt on my knees at the end of the day and it's there when I turn to Jared in the morning wanting to be touched.

And there is more of it, no doubt, making its way onto me now as I look back at Honk lumbering to the centre of the ice and hear a crack. Liquid appears from below the thin surface as her weight, and the heat from her feet, fractures the hardness below. I flinch, but she doesn't plummet; sinking slowly, gracefully instead – as if even an abrupt change in the very state of the matter under her toes doesn't ruffle her. She raises her head on a pipe-cleaner neck and plunges it into the new pool; grooming herself with a twist of her beak; splashing her back and in between her feathers in the way, I have learned, geese love to do.

I envy Honk's determination to start the morning with what she needs and her ability to roll with the changes. I stay here a little longer trying to harvest something from the sight of my white goose under the lightening sky. Though I'm trying to ignore and fight it in turns, part of me knows that – unlike her – I am plunging. This could go one way or the other.

It is there in every moment: concentration, pressure – from within – to: stop falling. Hold steady as the sun tracks east to west like this beautiful bird with her ridiculous name. And to stretch out and up too despite the fear. These instructions run through my head as the minutes pull a little further from the dark. I am coaching myself for what feels like it might be coming. I'm apprehensive, yes, but there's also a strange desire for the strength that might be released when everything starts to melt. Muddy water of oscillating atoms, ice bonds loosened by radiation, and kinetic energy splashed on to my wind-reddened cheeks like a tonic.

# BEGINNINGS

B reakfast time and I am readying the children for school, when Arthur throws up his Weetabix into my instinctively cupped hands. I drip his sick on the floor as I rush to the bin, the sink, find a bowl, kitchen roll and a glass of water; all the while watching to see if there is more to come. His little face registers a mixture of confusion and upset as I cuddle him and notice that the tender feeling of wanting to help is fusing with rising panic as the reality of a missed eight hours at my desk hits.

I keep up the comforting patter, but thoughts begin sliding out from the sides of my brain. Slowly at first and then faster and faster; a mechanised spitting of competing ideas that instantly go to war with each other. Worry, love and care are all there, but there's annoyance too and the guilt at feeling it. Then my plans, tasks and to-do lists start to intrude, with each thought spawning at least two more. If I don't send this email, plant this seed, finish this work, order this part, then I won't be able to do something else tomorrow; the impact of today travelling down a chain that stretches way into the future.

I am becoming familiarly lost in this paralysing interior pile-on. This feeling has been there in one form or another for as long as I can remember, but the intensity and frequency have really upped of late. With the smallest of triggers, I'm out of control and afraid but can't pin down what's frightening me or remember why it matters.

Arthur is sitting on a towel, bowl on lap and tissues close by. As I turn to wash my hands again – the smell of sick stuck to my skin – I try to reel in my mind like a lifebelt. It's an effort that's made just about possible by his little wobbly lip and the mother I want to be for him. I am trying so hard to stop this Tuesday morning unravelling and the thought of how much this matters threatens to make everything come apart faster. But I clench my jaw and tell myself not to go there, that I have time to take a day off, that everything can wait and that Arthur is the most important thing, which of course he is. As I stroke his hair and notice that his ears are dirty, I remind myself that I've changed my life over the past two years and work isn't quite the pressurised vessel it once was. This is not a disaster; I've just programmed myself to think it is.

The trouble is I know, despite this scaling back, that the stock feeling of running red-faced behind my life, reaching out to grab it as it keeps sprinting away, has not disappeared. I am very aware that this year is the year I am supposed to be back to my old capable self after the long, bad patch I've been in since we moved here. I should be feeling better, yet maybe, almost definitely, what I am feeling is worse.

Forty minutes later and the colour is coming back to Arthur's cheeks, he's eaten, has a normal temperature and seems absolutely fine. Typical. I am still working hard to keep myself in check, so I pull him on to my knee for a story that will absorb us both. As he leans in, ready to listen, his back presses against the bulges in my clothes and I push him away quickly before he crushes the eggs I'd forgotten about. Like a mad, ripped-dungaree-wearing magician I pull a duck egg, two hens' eggs and, the *pièce de résistance*, a bright white goose egg out of my pockets. Weighing over 200 grams, more than three times the size of a chicken's egg, the goose egg completely fills my palm with its smooth, ceramic-like surface and pleasing pointedness. We are not goose experts by any stretch. I'm not even certain how many males and females we have and, as this is our first year with laying geese, every egg is a thrill. Arthur and I gaze at it for a while and I think that finding a goose egg in my pocket on a difficult morning is exactly the kind of fairy-tale thing I want my life to be made from.

I nearly start the fable of the goose who laid the golden egg, but stop after 'once upon a time'. I'd like to keep believing that one day I might actually find a golden egg glinting at me from the straw-sided, feather-lined nest that the geese have built in the shonky field shelter we bought off eBay. I won't make it a myth just yet. I think again and as I do we are silent, our eyes on Jared waving as he passes the window leaving for work. When he's gone my eyes remain focused on the plot, scanning the veg patch, the fruit cage and then our field. I am looking for a story out there, trying hard to keep on the level by noticing every blade of grass and wind-cracked twig, willing on the leaves and breaking the quiet to point out a shimmer of catkins on the hazel copse.

I am holding Arthur close, for my comfort as much as his, when a local farm cat appears, being chased – yet again – by our ducks and goats. We watch the scene together, laughing as they noisily round up the cat, who flees over the fence in a shriek of black fur, into Victor's field where he spooks the ponies. I can't really hear the drum of their hooves galloping off into the distance from here, but my brain knows the sound well enough to colour it in. I heard it almost every day as a teenager coming from the fields behind the yard as we pitched in to keep the fees down at the Black Country stables by the motorway bridge. I heard it again when I rode through the nearby valley, flanked by housing estates and busy roads, but still a quiet, old place that calmed me. And so I can hear it now even with the windows shut: an urgence of percussion fading to nothing as the horses disappear down the hill.

Our plot sits at the top of this gentle hill that then tips down southwards for a mile towards the village school. When I first stood here I clapped my hands and grinned: a classic English countryside scene with fields, hedges, small patches of woodland and the occasional faraway cottage. This, I thought, is what I need to see every day. Lately though, when I look into that distance, I think I see other things too: indistinct but intriguing. They are beginning to appear across this February morning, as if a film were being projected on a too-bright day. This unknown scene is being offered to me gently at a time when most things feel as though they are flung at my head. I take it – this shadowy mystery – and try to find the very start of its story.

Once upon a time (I say again in a lower voice), a really long time ago, everything we can see out of this window would have been different. Let's go there, Arthur: back to when all this – I pause as one of the shimmers becomes more distinct – was the middle of a huge forest that covered everything for hundreds of miles. We are inside it now – and it feels as though it has been here forever: a density of trees growing undisturbed for hundreds of thousands of years. Thick trunks stretch way up, further than we can see, and little ones are dotted in between, sprung up when others, tired of holding their branches, blew over. My senses adjust and everything becomes clearer still: it is dark in here, isn't it, Arthur? A nice darkness, like a soft blanket. It's not exactly quiet – quite noisy in fact – but the sounds are dampened. The trees – so many – take some of the rustles, shrieks and the noise of the wind into their bark. Can you hear it? I turn to him – his gaze matching mine, holding my hand – and he nods solemnly.

This world grows out of the clay and somewhere inside me. We scan the scene for animals, birds and insects and it takes a little time because they are shy, though they don't yet know to be afraid of humans. What animals do you think live here? I ask. Rabbits, he guesses, worms? Yes, I say, lots of worms and lots of rabbits and foxes too. Red squirrels, dormice, earwigs, voles and owls – can you see any? We try, looking up towards where the sky should be and our eyes are so sharp when turned on this forest that we see them all. There's a nightjar sitting on her ground nest of two mottled eggs and we watch her blending in with the moss and leaves of a clearing to the east, giving herself away only by tiny movements of her folded breast. A louder sound next: over there. A wild boar rooting for acorns and – a sharp inhalation as I see them – wolves, beyond the marshy part where willows are growing. My little companion's eyes widen – he can see it as clearly as me: a grey wolf, less translucent with every minute. We watch the pack, a couple of cubs playing in the dust, and then I spot something large and feline: black-tipped ears and a spotted back. Arthur, it's a lynx, a big cat, shhhh.

I think he's hunting – I tell him – there must be roe deer nearby and my son looks upset at the thought of jaws and claws meeting a fawn's spotted coat. I nod, understanding and feeling the same but, I say, the lynx's mate has kittens now. I point to where the queen is occupied with washing them, her rough tongue pressing against their fur. The new family are sheltered by huge roots and the curved trunk of what can only be the greatest and most ancient of trees. A few weeks ago – I know it somehow – the mother cat lay down here to give birth, choosing this spot to raise her family because it felt like a place of protection. We look on as she lies down again now and the babies start to feed. Arthur, I remind him gently, she needs to eat to make the milk. He thinks for a few seconds: 'The daddy will have to kill the deer then.' I put my arms around him a little tighter as the tom lynx passes right in front of us on his murderous, life-saving quest. We stare at the canopy above and this world that has become so real that it almost obscures the sight of our tabby cat slinking through the open field on her way to hunt mice.

Late morning and the sun is high enough to shine directly out-side the kitchen. We drag cushions outside to bask in the heat; remembering how nice it is when the air is warm not biting. The sky is spookily blue for February; not the usual determined cerulean of late winter but a soft summer-like periwinkle. It's odd to see bare branches against a shade of sky that is usually paired with leaves, and even the crocuses seem shocked as they open and spin like satellites to stalk the sun.

This is all Arthur really remembers. He is tuned into the outdoor world we have foisted on him; noticing the moon's daily changes and knowing how to hold a hen so her wings don't flap. I wonder if he will choose something similar for himself or if, later, he'll resent us for taking him away from the big city where his sister was born. Whether deciding between city and countryside life will even be something his generation gets to do. But I try not to project into the future today, to avoid stepping any closer to the internal agitation that would otherwise catch light.

We eat lunch early and then I open three envelopes that came in the post, happy to hear a promising rattle from within. Money is tight this year, but seeds are a cheap dopamine fix and I have a burgeoning addiction. A pound or two brings me an assurance of summer and buys my freedom from the sensory assault of the vegetable aisle. I love them and they are useful too – the perfect thing to keep my son occupied and me distracted without it all feeling like a waste of precious time.

I take out the packets excitedly and my enthusiasm spills over to Arthur who squeals and shakes them, doing a funny little dance that I join in with. I tear the paper, pour seeds carefully out onto our palms and we consider them reverently: the pepper's pale curve; the darker, chunkier teardrop of the squash. These are the essential ingredients for a magic spell I am about to cast, and I resist the instinct to rush on. I roll them between thumb and finger for a moment instead, hoping that their ugly-duckling promise of eventual swan will rub off.

We sow them together sitting on the grass, getting seed compost everywhere. After we've done all the vegetables we can,

Arthur agitates for more and as I can't face a tantrum today, I fetch the jumbled box of flower seeds, taking out the ones I am most excited about: Cosmos Rose Bon Bon. They sound like something Jamie Oliver would name a child, and as we push these brown new moons into the compost, I am pressed closer to the pink, double-petaled flowers of July. Arthur must feel this forward motion too, chatting happily about what else we will grow for him to eat this summer. Raspberries, sweetcorn, peas, strawberries, carrots, beans, sugar snaps, lettuce (but not rocket – too spicy), cucumbers, potatoes, mint, beetroot (but only if I cut it into ribbons) and, finally, even more raspberries. I'd better get to work.

The next morning, both children at school and I am typing quickly. I'm skittering from task to task as if each is an emergency and yet entirely meaningless. I feel frantic but then catch myself staring gormlessly out of the window. When I get up to do something, I've forgotten what it is by the time I reach the next room. Finally, I settle to reading the headlines, clicking through to a story on how yesterday was the hottest in February since records began.

It's confusing to look at the still-seductive blue sky and know there's something sinister about it. Everything feels easier when the air is balmy, the light streams in and no one needs a coat. Being reminded that this heat is a symptom of a climate lurching out of balance makes these simple joys feel like betrayals. I should be outside now working the plot to make our lives less reliant on things that make the temperature climb. But – the follow-up thought comes quickly – surely it's absurd to even try to meet a problem of this scale by growing a few broad beans?

I read on and discover that yesterday's temperature record was logged somewhere not usually famed for its good weather – the place where the idea for our smallholding life first germinated. While Arthur and I sowed our seeds, the thermometer in the Trawsgoed estate in Ceredigion, 8 miles east of Aberystwyth, West Wales, crept up to 20.6°C. Less than three years previously on working holiday we'd passed that very spot and it was near there, with the children in the background running happily towards a stream, that Jared and I first talked about making a shift to a more outdoor life.

I remember that summer as one continuous picnic in the August grass – the memory already heat-hazed. Days passed and we turned browner and our hair sun-bleached. The world of trains, laptops, cafes and shops drifted away down the River Teifi and out into the sea. After nearly ten years of marriage Jared and I smiled more easily again, laughed often, played with the children in the day and talked and loved at night as if we'd discovered a secret compartment in one another with something new within every morning. I recognised a difference in myself too – not completely novel but a return to something I'd lost and then forgotten.

Soothed by the sounds of crickets, we found the space for thoughts usually pushed aside by work, hoovering and getting to the childminder on time. Late-night fireside chats evolved into a plan that felt like the only logical response to where we and everything around us seemed to be headed. I don't know who said it first – the wish to capture some of the freedom and energy we felt then and bring it to our everyday lives. But there was a shared certainty, I know there was, a realisation that we didn't miss much of the life we were away from and that there was a lot to gain in beginning again a little closer to the earth. Both our heads swivelled in a new way when we passed solar panels, vegetable patches and farm 'for sale' signs. And, as the fire turned to embers and we turned our feelings into a vision, we found that we were also talking about our biggest fears. That this plan we had somehow segued into making contained the worries we had about our family, our town, our country and our world.

I was thirty-four that summer; the summer when we started to sketch the life we wanted. The summer that Jared and I stood up from the fireside at the same time, scooped up the sleeping children and set off into the night without a backward glance. The summer of 2016: the Brexit vote and the Trump presidential campaign. The summer when, belatedly, climate change and the wrecking of the natural world transitioned from abstract worries to active threats. The summer when I started to tense against a feeling that there was a point of confluence ahead for which we needed to be ready. When, after six years of riding the peaks and dips of my little brother's destructive life, I was still surprised to discover the latest drop and find myself pulled down towards it.

I was thirty-four that summer; the summer of striding out towards a life of open fields and sacks of corn. And I brought my kids, Jared, two cats, three jobs and a confused black hole of something pernicious but not yet acknowledged along for the ride.

This afternoon, as I walk to the shed to search for seed compost, I take note of the welcome extra light. The days have lengthened by a full two hours since the start of the month and once again today the last of the sunshine lands a few minutes later and a ray or two further on the daffodils. I can see the summer garden emerging from pots in the propagator and on windowsills. Sweet peas are snaking up canes and the garden peas that overwintered in a plastic pop-up tent are already beginning to reach the top of their short stakes. The extra light is helping me tick off some long-overdue chores and I finally feel a little satisfied as I take stock of the garden and my plans for its year to come.

I am trying to focus on the progress I have made and not let myself think about the rest of it: everything else still to be completed, and how I feel when it piles up. This is why I am looking for materials to sow yet another batch of seeds. The tangle of energy, certainty, ambition, thoughtfulness, panic, self-loathing, anger, confusion – and all the other things not found or named – is pulling me in too many directions. Over these impossible feelings I choose the anticipation of a speck of green against black every time. I want to watch each one become a seedling, noticing its cotyledons, the temporary first leaves, open as if trying above-ground existence on for size before making a commitment. It feels as if identifying the precise moment that photosynthesis takes over could make the expanse of my thoughts manageable.

Despite these efforts this wide internal space opens up now. I see Jared and I sharing a mouse-infested shithole of an East London flat with five others. Just-out-of-university friends, thrown together as we tried to persuade the others to move further out for more space and light. We fell in love in my messy bedroom looking out over the twenty-four-hour Beigel Bake on Bethnal Green Road. Six months later we decamped to a tiny flat overlooking a courtyard, then another with a sliver of out-door space in which we immediately, unnecessarily, installed two rabbits. The dense undergrowth of our current life is seeded in these beginnings. Getting married the week after my twenty-fifth birthday, a curly-haired baby at twenty-seven; always wanting to tangle ourselves up with where we lived. Jared keener to stay still for a while and me always restless and ready for the next thing.

The lurching strangeness that's becoming obvious in me must have been present in the soil throughout. And it's the same out there in the world – the leaves of this month's record-breaking heat having their moment of vernation one hundred, five hundred, perhaps a thousand years ago. Once I start, it is hard for me to stop this trail from broadening and dragging my thoughts to the uncertainties of the future; the demand for a response and solution following quickly behind.

I force my focus away from this searching scan, leaving the shed, tipping the compost I've found into a bucket and concentrating on some weeds I've just spotted. I kneel down to tug them out and my thoughts drop to where these roots in my hand have been growing. Since the forest told itself to me in a story I am looking anew at what's under my boots, doing research and trying to understand my land a little more. The need to pull things apart, to examine them and try to make sense of *now* by understanding *then* is not new to me. I don't know why I'm compelled to spend time I don't have on these investigations, but I have given in and jumped back as far as I can. The sun has risen and fallen 46 billion times on this clay since it was laid down 130 million years ago. The earth has stayed solid here as land that became sea that became land once more. Next, it hosted a warm, wet jungle that was ended by the suddenness of extinction and followed by almost endless winter.

The thickets that grew out of that eternity of cold became the prehistoric, prehuman forest of the Weald – the blanket of leaves I spun for Arthur is a true story. Bacteria, plants, birds, insects, reptiles and mammals emerged, altered and gradually disappeared. In time creatures whose descendants would eventually stand up and usher in the Holocene moved across its leaf-mould floor. And it was only in the forest's yesterday that the slow pace of everything picked up again. Its sucking soils and dense woodland had long acted as a barrier but by and by – a little over a thousand years ago – this place, untouched for longer than most, felt the force of permanent human inhabitation. Small clearings for pigs and tents multiplied and expanded; a tsunami of change had started. Axes met saplings and large trunks alike and eventually even the Mother Tree – the oldest in the forest, charged with protecting this woodland (with a sideline in sheltering generations of lynx) – was felled. Within three centuries, 500 square miles had shrunk to 150 and sunlight hit grass more often than leaf.

Coit Andred. Andredesleage. Andredes weald. I mutter these words as I pull more weeds up: the oldest names I can find for the forest that used to grow where my house now stands. Looking up, I see a clearing – sky, grass, meadow, field and a place to grow food – a felled forest reflected in our windows. The past doesn't seem keen to step aside here. The more I look and listen, the more clearly it comes into view. Holes in the canopy, paths through the darkness, the smell of woodsmoke and human voices getting louder, life getting better. Land that had just existed, just was – a place to walk across, scurry on, climb up or fly over – becoming somewhere that was owned. A safer world of hospitals, welfare state and life-saving vaccines, yet one where trees became the exception and fields – dotted with houses, bordered by roads – the rule.

By the time our family began our new life here, there were only 23 square miles of ancient forest left – half damaged and a quarter over-grazed. As we explored the nearby countryside, we had no idea that many of the villages, roads, shops and holiday cottages had names that came directly from those of the earliest settlers and the first clearings they made. I didn't realise, as I made a start on turning the soil with my fork, that I was digging myself into a past about which I had no idea.

Until I was thirty-four, and those warm weeks of summer in Wales did their work on me, I was convinced that I couldn't successfully cultivate so much as a layer of mould on jam. Our last home – a tall, narrow townhouse – had a garden that pleased and terrified me in equal amounts. I wanted our family to be out in it and for it to be lovely, but I didn't know where to start.

With no practical experience I was afraid of doing something wrong. How was I supposed to know what to feed and what to prune? When to water and when to stop? I worried that I would kill the plants and break the rules, not realising that they were my plants and my rules, so it didn't matter. I left it, concentrated on stopping the house from falling down, on working, looking after a toddler, then being pregnant again and having a new baby and tried not to look outside too much. Eventually, when the mess became oppressive, I paid my gardener friend Catherine to design and plant a low-maintenance bed and panic-turfed the rest.

Catherine's quiet work did pay off, despite my ignorant refusal to spend money on topsoil and compost. We had colour to look out on and a set of very comprehensive instructions, which meant I did remember to water – at least sometimes. Looking at those plants, noticing the buds form and turn to flowers and knowing I'd played a small part in that transformation, began to do something to me. I found I quite liked it out there, by myself, with only the hose and a ladybird for company.

One late-spring day I felt ready to take the next step and create something of my own. I remembered the impulse-purchase strawberry plants I'd picked up the previous year, found them looking half-dead at the bottom of the garden and planted them up in pride of place near the house. For the next two months I watered and fed those strawberries, tenderly snipping off brown leaves and evicting weeds. It was my first real taste of gardening and I was proud of how well it was going. By the time we set off for our summer in Ceredigion my soft fruits were in rude health and had quadrupled in size and Catherine, who was watering in our absence, promised to pay them special attention.

When we returned five weeks later, muddy, happy and ready to ditch town life for the call of the countryside, the strawberry plants were covered in reddening berries which we stood in a circle to admire. I felt the addictive sense of achievement and self-reliance that comes when the work of tending to a plant turns it into food and it seemed like a validation of the decisions we'd made while we were away. If I could rescue those almost-dead things, maybe I could do anything: raised beds, a walled garden, the Hanging Gardens of Babylon perhaps?

It was a week later that I realised I'd been scammed. The strawberry plants hadn't even had buds on them when we'd left for Wales and now they were covered in suspiciously ripe fruits. They looked almost like completely different specimens. And that is exactly what they were. The plants I'd bought twelve months previously would have died within a matter of weeks, because if you leave a plant, any plant, even a cactus, in a tiny plastic pot for a hot summer with no water it will shrivel to nothing. Then, when the spring rains water the wind-scattered seeds, a new crop will appear. A crop of weeds. Weeds that kind Catherine, because she had seen the beginnings of a gardener's glint in my eye, had dug up and secretly swapped for the garden centre's best strawberries.

My cheeks were as red as those berries when I realised what had happened, yet her ploy had worked. Though those fruits hadn't been mine, I knew that they could be. Six months later I was planting real strawberries in the old fruit cage of our new smallholding and realising that all I needed was a trowel and the desire to begin.

# FRAG MENT ATION

Over breakfast something small finally tips me off that ledge – the one I have been balancing on for quite some time. I hold the glass jar of coffee beans in both hands over my head and bring it towards the floor, via the top of my skull, with force. Under the instant landslip of adrenalin, I realise I've been waiting for this moment: secretly hoping it will come, rigid with the effort of preventing it. It's almost a relief to give in.

I'm hoping for blood: a gash and then a warm trickle that fills the creases of my forehead and drips rhythmically onto the kitchen's grey tiles. I am sick of trying to explain this with words that I can't get right and that everyone seems to ignore or forget. So today, in a split-second reaction, I decide I need to leave enough of a mark so that even when I wipe my eyes and pull myself together later, neither Jared nor I will be in any doubt that there is something very wrong.

Shelf, hand, forehead: the jar hits the floor, rolls and barely even chips. Jared looks shocked, angry even, but not full of realisation. I put my fingers to my forehead and they come away clean. There is only a dull ache and a stupid little egg on my hairline and it is not enough. So then something else: better, louder, more painful. Four delicate pottery cups, handmade to look and feel like wood and just the right size in my hand. A decade-old gift from my mother that I had wrapped and unwrapped each time we moved, relieved that they hadn't accidentally cracked along the way. Today I throw them all at my favourite piece of furniture, an antique dresser that displays our family treasures. Photographs of the children, our wedding; 1940s Martini glasses collected for Jared's birthday; a card hand-printed by a friend reading only, 'Peace'.

The dresser door smashes as the first cup hits. The second takes out the objects within. I barely register the third and the fourth meeting my palm and leaving again a split second later. But they do. Last of all, perhaps for completeness, I pick up the matching jug. The final piece of a set that had been chosen with care and given with love. I lob it like a grenade.

There is rubble when a person explodes. Glass everywhere mixed with shards of ochre, duck-egg blue, deep black and winter green. A tasteful rainbow of destruction underscored by little breathy sounds of shock and a cockerel crowing outside.

I am blank. I hear static noise, the ends of my fingers tingle but the whirling anger and panic has finally gone. My feelings are in pieces on the floor and that's fine. I sure as hell don't want to stick them back together again.

I stand there, numbly, waiting for the fall-out, the reaction, the judgement to be handed down. For Jared's strong arm to land on mine and march me out and, in a shout-whisper, tell me that I can't behave like this ever again. Or maybe to be scooped up and put to bed with an ice pack and instruction to sleep. Perhaps he'll insist I cancel my day of meetings and get some help right now. I don't know what it is that I want from him, apart from a response.

There is only silence. Jared is rooted to the spot and the children – who have just appeared – are looking on with huge eyes and pale faces. Centuries pass while I wait for him to meet me here in this awful place – a corner with no way out where everything is dangerous and broken, including me. I am not even trying to hide it any more. But I just stand there in the quiet, feeling a hot wind blowing across the fields behind me, hitting my back and trying to push me closer to the danger. No one is coming and the all-aloneness of the moment becomes more obvious with every second that passes. It is just me, in the dark, and I can't bear it.

And so eventually I stop waiting. I pour myself a cup of coffee, open the cutlery drawer and use a knife to spread peanut butter onto bread. I collect my bag and keys – an odd calm bolted over the adrenalin rush – and pick through the room hearing fragments hit the walls as my feet send them skittering. As I leave to catch my train, turning back with a shout of 'have a nice day', there is no reply. All three of them look at me as though I've lost my mind.

Which of course I have.

My day of work goes well despite the preface. I'm good at this stuff, whatever the run-up. I put the morning away somewhere I can't reach – pushing it further as the train takes me to the city. When I'm there, meeting people, I don't have time or space in my head for anything else. As well as actively reminding myself to listen to what's being said, I am also instinctively taking in the bigger picture and decoding the dynamics of the room. My antennae are tuned to the unspoken things that matter: a shadow under someone's eyes; a micro-glance at the water bottle as they balance thirst with not wanting to interrupt; the way they hold themselves taut at the mention of someone else's pregnancy. I am at capacity with the effort of putting all this together, making a formula and turning myself into its answer, so it's only when I'm walking through the streetlamp-lit station car park that this morning becomes real again and I start to feel a bit sick.

At home there's the kind of quiet that tells me the children are in bed; a hush that highlights how nervous I am about facing the intervention to come. The house is spotless – tidier than when I left. All traces of what happened earlier have been removed and if you didn't know the things that were here before I destroyed them you wouldn't miss their presence.

'Hi,' I say with studied casualness as I walk towards Jared in the kitchen. 'Hello,' he replies; a little wary perhaps, but not hostile. He asks about my day and so I tell him: the meetings, the promise of something exciting, the annoying email. He offers things up from his day and, though we are stilted, a light remark and half-laugh seem to go over okay. It takes me half an hour of this to realise that the conversation I have been expecting is not on the cards. If I don't bring up what took place over breakfast it might never be mentioned again. I know I should do it: be brave, apologise, try to explain why it happened and how I am going to make sure it doesn't again. Make a plan to fix this, fix us, fix me; but I have been doing versions of this for a while now and I don't have anything left.

So instead, after eating and watching something I can't remember, I lie on the precipitous edge of our bed and think about that summer day three years ago when we started to talk about the future and ended up, five months later, on this plot. Excitement, shared purpose and a plan – I tell myself that these things propelled us, but I wonder if there's another story that's also true. In this story it's all my fault. I'm the impulsive one who pushes harder and has it all scoped out with the answers to every possible question prepared in advance. A story in which I have good ideas and a drive to bring them to life so Jared plays along. And somehow our new life becomes my project – though I'd never have set out to do it alone. I am unchecked and leading the charge and so I take the responsibility, the credit and the blame as we move apart instead of together. Maybe that's the real story. Maybe not.

I want to understand what went wrong, but the past has split and I don't know which version is the original. I don't know whether it's normal that we are not talking about the way I shattered into sharp pieces this morning. I can't tell if something big and bad happened today or if everything is fine. There are so many possible scenarios and I have tried to make a plan for each, just in case. But these responses, every one given an equal weighting, don't resolve anything and instead become staccato fragments arguing in my head. Sleep will shush them for a bit though and so I try to get to it by focusing on the only thing that feels certain: the ceiling, the air and, above it all, the expanding infinity of space – which has nothing on the distance between my husband and me on the mattress tonight.

The next morning we have carried on ignoring everything that matters. Still, I took the last bit of bravery I must have stored in my little toe for such emergencies, and began the one piece of repair work I knew I had to do: apologising to the children. I pulled them close to show them that my knee was still a safe place. I told them that I understood how they might feel because when I was a little girl it frightened and upset me when people would volcano their feelings seemingly out of nowhere. I said that I was so sorry and that I would do better. They appeared to be fine, taking it all as read, relaxing into my arms and, now they are at school, I try to reassure myself that it's the apology that counts, that all I have to be is good enough as often as I can. But this was not good enough and I know it.

I have already written off today's work. I'm drained and wired and at my desk reading the news. 'Apocalyptic,' says a BBC story about the burning of Saddleworth Moor, one of the largest-ever moorland fires that sparked up a week ago as my little son and I sowed seeds. In North Yorkshire, fanned by one of the driest and warmest Februarys on record, two-metre-high flames moved swiftly across the early-spring land as if it were the end of a long Californian summer. As if it were the end of the world as we know it.

I read eyewitness reports and follow the hashtags that fly out from the story taking hold online like sparks on dry bilberry and cross-leaved heath. We have spent the past two hundred years sucking the life out of the moors and now they are dying. But as our future is tangled up with these great carbon sinks, people are trying to save them. Last year they were down on their knees planting a million sphagnum moss plants to help restore this land to a featherbed bog of moss and cottongrasses.

Much of their effort has just burned away to nothing – the peat giving up its treasure and pumping carbon into the atmosphere at a rate of thousands of tonnes per hectare. 'Apocalyptic,' says the story, but I wonder at what point 'apocalyptic' becomes 'apocalypse' and whether it's possible to catch the moment before the adjective becomes that definitive, deafening noun.

I close my eyes against the weight of these facts and thoughts but since they are inside my skull this only intensifies the smash of jagged edges as they collide. I needed a calming morning, but instead I am keyed up and my thoughts are flinging themselves against a heavy tiredness made of yesterday and its yesterdays. I should be doing more, be out on the moor on my knees planting sphagnum moss until the trowel handle bursts the blisters on my hands and begins to make new ones beneath. My heart is full of more and moor, but I can't calibrate whether I am twiddling my thumbs and being lazy or if I am already at capacity. The space in my head is a wild moorland filling with a snarl of leaves, stems and roots that make an ecosystem of thoughts and plans. Golden plover, dunline and short-eared owls take turns to run and fly and dive across it in random bursts until I lose my place and everything starts to spin.

Zooming out and seeing many things at once, piecing them together to spot patterns and construct self-supporting plans that capture everything from the micro to the macro – this used to be my biggest strength. But I have managed to turn my brain against itself and this hasn't been true for a while. Perhaps I was denser, heavier and at less risk of tumbling out of control in the past. Maybe I weighted myself down before: used pegs and a rope or wore a protective suit made of something like sphagnum moss, a soft, acidic layer to stop everything evaporating. Yet I've lost sight of whatever it was that used to keep me safe and, until I discover what to lash myself tightly to, I can't go further into the moorland, not after yesterday. Feeling guilty, I close the

browser and go outside to look for something more solid than
peat bog on which to rest.

I walk towards and along the boundary, tracing its route from
roadside to field. I notice trunks at first; different widths and
barks, all lined up either side of the fence like security guards
and it's then that the extraordinary span of the largest oak's
canopy registers. I look up at it for a while, relaxing in my insig-
nificance. The shade I cast wouldn't shelter more than a handful
of bluebells but this tree could shadow a planet. I know to be
reassured by its size somehow and maybe, yes – another idea –
its age. I should try to work out how old it is.

A quick Google sends me looking for the dressmaker's
tape measure inherited with my grandmother's sewing box. I
approach the fattest oak trunk and put my palms on it, focusing on the way the bark feels under my nervy hands. It is quiet
and still under the canopy with decades of leaves that make it
slippery underfoot. As I try to get the tape measure around the
trunk I lose my balance and brace for another fall, but the catch
I've been hoping for comes this time. Spinning to see who held
out their arms I find only a laurel branch.

I measure in sections and make slapdash calculations on my
phone until an answer appears: 335. The numbers settle slowly
on me: dandelion seeds following the breeze's path until they
finally land. This tree I have been ignoring is 335 years old. It
has been growing here since 1684.

I take it, this direct link between now and then that feels
like a way out, and travel from the sharp pieces of the present

day to when everything came together perfectly in the moments after an acorn fell.

❦

1684 has become a gap that I need to fill so, inside and on the internet, I click, skim, highlight and soon have twenty tabs open. Unlike everywhere else, this feels like a safe place to get lost. Nothing in 1684 can be my fault and no one from 335 years ago is expecting my call. I feel free to spread out across the year and, as I do, I find scraps that might fit together if I find enough of them.

The acorn's taproot began to tunnel under the earth as Charles II lived out his final year as king. England struggled through the coldest winter in living memory and a 'little ice age' gripped the land. South from here, in Romney Marsh, the anopheles mosquitos' malarial bites were stopped for a short while by the chill and the sea froze for two miles towards France. In Cambridge, Isaac Newton kept warm by the fire and scratched away with a goose-feather quill to prove and explain much of what had seemed mystical until then: tides, comets, equinoxes, eclipses and the relationship between the way objects moved here on earth and in the big skies above. 'My' acorn hit the clay below, as every acorn has ever done, because of these laws of gravity. But perhaps it was the first to fall in a time when we knew what force was holding us, and it, to the earth's surface.

More changed that year than our understanding of the invisible. The thirty-eight babies baptised in our village church would grow up with a chance to go to the school that had just opened. The village had gained a blacksmith: John Illenden, who felt the heat of the new forge's fire on his face every day as our oak grew to a sapling. Snow, ague, celestial bodies, water splashing on thirty-eight foreheads and a horseshoe glowing white in the heat – single random stitches, the start of a tapestry, but fabric in the making nonetheless.

I am finding this new occupation as comforting and full of secrets as the seeds I love to sow and am about to click again when I look at the clock and see the day has all but gone. As I drive to school I see that anonymous houses, trees, hedges and fields have become individual question marks. Histories that want to bump up against mine. But right now I have to leave all this in the glovebox, open my arms to the children and hope they will find me a little steadier this afternoon because I've lashed myself to the ground with gravity and acorns.

March has nearly passed and the daffodil show is at its peak. Thick ribbons of lemon-white and yolk-orange enclose the trees in the orchard. Even on difficult days, it is impossible to ignore them in this volume. As if the plot has been building to this all winter and is releasing the stored energy in a six-week-long circular breath.

I am picking them now – daffodils for the kitchen table and others to give to a friend who needs cheering. With each diagonal snip of a stem I have the feeling of settling into a pre-made mould. My fingers on the scissors, my boots pressing into the squelchy grass and the sap leaving viscous smears on my gloves. Another woman planted the bulbs that became these flowers.

Every spring she would cut them as I am and bring them in to die vibrantly in the warm air.

As I take my bunch inside, I see birds overhead, returning from winter travels. Their size, the delicate point of their beaks and contrasting scope of their wings all instant clues that these are geese. I hear their echoing bugle sound now and know it well. It feels a worthwhile thing to have learned here alongside the discovery that there are many more than four seasons. I used to think that spring arrived overnight with blossom and summer showed up just as suddenly on the first hot day. I would know autumn only when I couldn't see the pavement for leaves and winter would wait until, pulling on my hat, I realised the branches were bare. Now the year's cycle turns continuously because I am lucky to be so close to its daily signals.

After lunch I remember to go to the shed to check for more plant labels. I find a few on a shelf and then rummage on the floor, lifting the boxes where the few dahlias I still need to pot up have spent the winter. As I move another box, the shed fills with the noise of broken china – no dahlias in this one after all. I open it and find shards of pottery, glass and broken mementos: the tangible results of my destructive morning. I run my fingers over them and imagine Jared sweeping the floor, holding the dustpan and not knowing what to do with its contents apart from find a dark corner to hide them in. The memory of that morning, how quickly I put it away and how busy I have been distracting myself from it, are all instantly and horribly clear.

We still haven't talked about it and though I haven't lobbed any more heirlooms at the furniture, neither have I found a way to be sure I won't do this or something worse some day soon. I press my fingertips into the pointed shards, which have now joined with the other sharp things lurking at the back of my skull recently. Painful, piercing thoughts, memories and worries; intrusions of scissors and kitchen knives.

I snatch my hand away, carry the box to the bin and throw the whole thing in, walking away from the crash it makes, determined to tackle the last dahlias and move on. I inspect each tuber for rot, use a lino-cutting knife to divide the largest into three smaller sections and then plant each out into a good-sized pot where I'll grow it on, sheltered until the frost has passed and the slugs have lost interest. As I work my way through this job, fiddly as it is, I can't shut out the replay of the insistent sound of breaking that box contained. I've forgotten the specifics of why I was so angry and desperate that morning, but the feeling has remained under my various pretences. At last, because it is about to consume me anyway, I stop pulling away.

There have been plenty of what looked and felt like good days this month. Everyone is desperate to believe in them and in the steady springtime version of me: smiling, efficient and calm. I have tried to convince myself that she is real and that the woman who exploded was a blip. Except she wasn't. There have been more and more blips and they are getting closer together; their trigger more sensitive. However hard I have been pretending, I know that I smashed things I loved so that we could no longer ignore that I am coming away from myself. Now with my eyes closed I make myself face. I go back to the beginning of the month when I felt like my only option was to break my way

out: kitchen tiles becoming dry ground, fields behind me and a
hot wind blowing all around.

I am alone and have been walking in this dark forest for months.
There's a heavy bag tugging at my shoulder which I reach inside
to pull out red-flagged markers that I place in the ground. Each
pinpoints something vital: danger, the easiest route, a dead end,
the way home.

I spot someone coming towards me as the gloom deepens. I
know who it is and I am overjoyed to discover that I am not alone
any more. He has remembered and wants to see me enough to
brave the forest at night. My markers must have worked and
this makes the welt the bag's strap has left on my shoulder feel
worth it. I wave excitedly, but he hasn't noticed me after all and
is instead heading for one of the danger zones. I leap over rabbit
holes as I run to him, grabbing his arm and spinning him around
before he can go too far. I am about to kiss him but then I see
his own bag full of my markers. He's been pulling them up, as
if I hadn't told him what they were and why I needed them. He
was collecting them for me, he says, bristling at the sight of my
face. He was doing it to help, he knows I like things to be tidy. I'd
asked him to walk with me in the forest, and he's here now, isn't
he? And yet I am shouting at him and crying – he doesn't understand. 'We won't get lost on the way back!' he insists, refusing
to believe in the dangers and dismissing my need for a specific
route, because he doesn't want to know how weak I am.

A sharp spike pushes into my skin all of a sudden. We've
been moving as we talked and I'm backed into a corner now,

a barbed-wire fence behind and no good way out. He tells me to go round, to climb over, go under or just forget about it. Everything is fine. But he has no idea how it works in the forest at night and the more I tell him the less he hears because he doesn't want to know. I try and try. I am like a coiled spring and he a blank wall. I bounce off him and fall on the floor with a clatter, rolling from side to side by myself until I am compelled to try again.

This has to end; I have to leave this place. There is one option open to me, but I don't want to do it because, though I'll find myself freed, it will send me in deeper next time. I make a final attempt to get him to see all of this through my eyes and take my hand, but it comes out angry and garbled and he pulls away further feeling slapped and small. I have no choice. I pick up a rock and then another and another, lobbing them at the trunks of the trees until their bark splinters. I pick through the wreckage while he stands in silence. I don't know what he is thinking, but he doesn't follow me. He didn't heed the warning, understand the forest or believe in what was about to consume me, but he definitely sees the mess I've made trying to escape.

# A CANDLE IN THE DARK

I've been waiting for the reminder that has just appeared on my phone screen instructing me to 'CANDLE THE EGGS'. Not that I needed it – I've been excited to check on their progress since I set sixteen eggs to hatch a week ago. I turn the incubator off and load the warm shapes into boxes as quickly and carefully as I can, imagining the gory scene if one drops to the floor. In a darkened room I hold each up to the bright light of a special torch bought in our first summer here. Several hatches on and I know more or less what I'm looking for. I no longer need to watch fifty-seven YouTube videos to be sure that these red veins spidering out from a little blob are good news.

There are, as always, some duds and disappointments. Two glow clear apart from the circle of yolk, suggesting they were never fertilised. There are a few where life must have started and then stopped, its brief flicker captured by a telltale ring – a blood line – and little pieces of tissue floating in the albumen catching the light of the torch like dust in sunshine. I discard these to protect the developing eggs from bacteria, but as I've learned to set more than I need, I'm feeling good about my chances of a clutch of chicks in two weeks' time.

I put the promising eggs back into the warmth, rotating them as a mother hen would, top up the water and check that the temperature is beginning to rise to 37.5°C. There are plenty of flashy incubators that would do all this for me and make a better job of it, no doubt, yet even if I could afford one, I like being tied tightly to our smallholding for the three weeks this process takes. The precision and regularity of the tasks is helping to contain me – soft little walls on which to lean several times a day. The expansions and contractions of last month need to be dealt with, but I am still putting it off in blossom-scented, downy-soft ways. Each laborious turn of an egg is really a distraction and an act of faith: there will be a chick in my pocket, new breeds for the flock and rainbow colours in my eggboxes again.

As the air started to warm in our initial February here, I began to take wobbly, introductory steps as a smallholder. Within weeks I had a dedicated bookshelf of second-hand tomes that promised to teach me about vegetable gardening, companion planting, how to grow without any back-breaking, microcosm-disturbing digging, about living self-sufficiently on an acre, keeping poultry and more. I borrowed Sofya's colouring pencils to sketch my vegetable-planting plans: orange for carrots, green for cabbage, yellow for the onions planted around the edges to keep the root fly off the scent.

I sowed my first-ever seeds in trays (badly) and covered the beds with sheets of blue plastic found in the shed, on a hunch that it was the right thing to do. By March I'd decided it was time for animals and brought home three ducklings who panicked every time they saw us and I panicked right back at them when they refused to go under the heat lamp. It would be twenty weeks before we could poach duck eggs for breakfast, but as that first spring warmed to summer we started to eat other produce. I cut asparagus with a sharp knife and, though I didn't yet know what to do with the flower beds or how to keep on top of the weeding or the pruning, the garden rewarded me anyway with roses, clematis and irises.

We were hardly ever indoors: a family delighting in our surroundings and how they changed us. Sending Sofya out to gather kindling and marvelling at her strength when she came back dragging a twelve-foot branch. Watching Arthur strolling easily in the field pointing at our neighbour's sheep and lambs, the sun sliding sideways casting a halo around his head. Always out doing, digging and making plans. Idyllic.

Two years on, as I look at this time, I see another, less idyllic, side to all that doing: the beginning of many things becoming wound up far too tightly. In the blur that those twelve months have become, it is hard to tease out the moment I went wrong. Perhaps it was a few weeks in when, even though I was always working at something – ten things – I was falling behind. I didn't know what was needed or how to do it. I hadn't got any shortcuts or handed-down secrets and neither Jared nor I could easily knock up a frame for beans to climb up or fix a broken shed door. Just doing some of it badly was a full-time job with lots of googling and swearing. Yet the harder it felt, the more vigorously I attacked the project, never stopping to listen to myself or the sound of the marsh frogs in the distance.

Maybe it went wrong a couple of months later, when I had to face up to a big pile of work projects that stretched me in all directions. On top of this I needed to get the builders in: we didn't have heating, the hot-water system was about to give out – I didn't think it could wait. Or perhaps it was not a single moment but the creeping, not-really-realised feeling that Jared, while happy to help when I asked, wasn't diving in too. I worried that I had taken over and elbowed him out, or that he was regretting the move. It became a small sore spot that I tried to heal by showing him that the dream we'd planned together was coming true. I attempted to do everything, all at once, and tried to do it bigger and better in the hopes he would join back in. We became I: I, I, I, I, I. And soon I, I, I, I, I started to feel quite different from ever before.

I did not listen to the danger signs: tight in the chest, irritable, unable to sit still, to speak slowly or make decisions. I did not go gently into my garden and potter, but planned and executed a full-scale vegetable assault. I wanted to expand the growing space of our already-large plot with the addition of three more raised beds. This felt very urgent because when someone mentioned a vegetable or herb I wasn't growing I felt a very real sick-making stab of anxiety that I couldn't rationalise. If I grew everything though and grew it well, this would not be a problem. If I had enough growing space I might get rid of the sensation of running up an escalator moving rapidly in the opposite direction – knees, arms and heart pumping away furiously only to find that I remained static or had slipped back. I had tied bits of myself, of us, to this project and so I couldn't stop, I couldn't fail. Instead of resting when I felt shredded, I hammered the nails into the wooden frames for the new beds myself in the early morning. Bang. Bang. Bang. The sound of someone increasingly out of control but doing a very good impression of having the time of her life. The beat of Jared's feet stepping back from what we had started together.

The work dial turned up then in tandem with the summer's extra watering. Everything was enormous and urgent and I would start each day almost high with it, a feeling that increased as the hours passed: adrenalin bursts accompanying every pulse. I wrote very long lists, thwacking the side of my skull with my hand regularly to try to keep my brain on the task in hand, as it ran off like spilt toadflax seeds across the kitchen counter.

By midsummer I was faster, clumsier and planning out my day from 6 a.m. to 10 p.m. in thirty-minute increments. At night I would hit a physical wall of exhaustion and thick sleep would come almost instantly. In the cracks of time between my day job running a small but busy charity, writing a book, speaking at events, sowing, growing, digging, feeding, mucking out, fixing and reading about ground-source heat pumps, I tried to be a friend, a good mother and a wife, but found only impatience and an urge to be left alone. So, I took up running, tried to fit in hobbies I had enjoyed in the past – as if what I needed was more not less. I was spinning through space; except space was my desk, the plot and the inside of my head. I was both stuck in it and I was it. Gravity – that constant – seemed to have given up on me. To anyone outside, to Jared most of the time and often to myself, all that was visible was the excited smile of an ambitious woman. I didn't know which of these stories was true: the idyll, the person wrestling with a full but rewarding life, or the feeling of complete disintegration.

I kept going though. I hid it really well: how I was feeling, the loneliness, the fear and the truth that this land no longer felt like a sanctuary but a battleground. By the end of that first summer I had finished my book, grown courgettes and the raised beds were full of ripening peppers – our first chicks pecking in the dirt beside them. I had achieved something else too, though I didn't know it then. I had broken something in my brain and, to this April day two years on, I haven't been able to work out what it was or how to mend it.

Eighteen days after I put the eggs in the incubator I am candling them a final time. The children watch as I hold each up to the torch. There is less to see now – the chicks fill almost all of the space but the air cell is visible in each, ready for the chick that we hope will break through the membrane into this place of transition. It will rest then before, in frantic bursts, bashing a pointed egg tooth – a tiny temporary axe on the end of its beak – against the calcium carbonate wall. Sometimes the chicks don't break through before the oxygen runs out. As advised, I try not to intervene for fear of killing a healthy chick, but it is hard to wait. I have peeled away the shell too late before and seen the little creature a millimetre or two from life, all wet and tucked up – a kernel of not to be. Today I increase the machine's humidity to soften these shells and give this batch the best chance of making it out.

'Why is it called candling, Mummy?' asks Sofya as we leave the room. The answer is obvious, so I give it unthinkingly, hurrying the afternoon towards bedtime with a confident explanation that people used an actual candle for this task before they had electricity. The children are satisfied with this and move on instantly but as I run them a bath I find that I am not.

The blue egg is unzipping – as if worked at by some internal tin opener. The top of the shell is almost off and, as I watch it alone with the heavy duvet of midnight around me, hours of slow progress are over in a moment and confinement becomes liberty. The shell is empty and broken, instantly meaningless as it wobbles from side to side on the floor of the incubator. There's not a trace of yolk or white inside to tell of the journey that has just ended. The chick isn't the fluffy handful of greetings cards – but a smear of a creature, its down oiled with the contents of the egg and its new body pushed to breaking point by the effort of escape. I wonder if it has already forgotten becoming something from nothing and then smashing its way out. I let it rest while the other chicks, still in the invisible prelude to being, try to emerge from a fracture line into the night.

I've been trying to work since 9 a.m. but am distracted. When the children rushed in at seven this morning to tell us two chicks had hatched and two more were giving it a good go, I pretended I hadn't been up since five and seen it all already. Now I want to check on progress once more, but have promised myself I will wait until lunchtime. However, I've made no such promise about my next piece of research – the history of 'candling' – which has been tapping on the inside of my skull for the past three days. My fingers type out the odd-sounding word and I'm soon looking at instructional videos, over-priced hatching tools and reading the occasional mention of a lit wick and a steady hand. I find no indication of when a person first picked up an egg, held it to a flame and thought, this could be a handy way to count my chickens before they've hatched. Yet there must be a history to this process. Some confluence of curiosity, economic necessity and ingenuity that compelled someone to try and see what was inside. The knowledge must have been passed along too – horizontally and vertically across place and time. So why hasn't that left a trail I can find?

I have my teeth around this now, so I keep going, looking through the elderly but new-to-me books about small farms and Kent piled on my desk – the results of late-night eBay sessions. There are three surveys of the county, written by a series of men who seem to have jumped up every hundred years or so with an urge to walk across south-east England for no other reason than they could. I am glad they did because the books made from their journeys have already been useful, but they make me feel uncomfortable too: all headlong trample without pausing to listen and look – I can see myself in them.

William Lambarde is the earliest surveyor in my pile, setting off on his *Perambulation of Kent* in 1570 when the country was still divided into the Saxons' 'hundreds' – parcels of land that could theoretically sustain a hundred households. Lambarde writes about our village, some of its fine families and the forest of the Weald. Some men, he says, insist that the untouched forest of old began in one place, while others are sure it was quite another. He doesn't buy their certainty though and exhorts that a man – always a man – 'may more reasonably mainteine, that there is no Weald at all, than certainly pronounce, either where it beginneth, or maketh an end'. Lambarde's forest is one that can never be known and I feel as if he is telling me not to bother even trying to bring the plot's trees and what was or wasn't candled under them more sharply in to view. I put his book down.

Nearly two hundred years later Edward Hasted wrote his own history. According to him Coit Andred – the forest of the great ford, to translate the ancient Britons – was a waste desert and wilderness for many an unpeopled year. Hasted gives me something good – the words for the first clearings: dens, denberies, wealdberies – and then he walks on, travelling north along our road to hand the writing baton to William Henry Ireland. Ireland, like the two fellows before him, has pages and pages on kings and lords, dates and rents, but not a syllable on the poultry farmers who must have been here. Not a sentence about candling. It is not reasonable to be this irked by the long dead, but I am. These men were trying to capture this land and tip it onto the page, but as they put their boots on the ground they seem to have noticed only whether it was wet or dry, wooded or open, fertile or bordering a stream. They didn't seem to wonder enough about the lives, stories and secrets they might have discovered if they weren't walking so fast.

Annoyed with the books and myself, I get up to check on the eggs. Another chick is out and it shouldn't be long before the fourth joins its sort-of-siblings. Watching the first bird fluffing up while other little beaks work away inside their shells soothes me a little. I don't even know exactly what it is I'm looking for at my desk, but I know it is not going to be found in these assertive accounts. I look at the egg rocking as the chick pushes on and think about the wives of these men. What it was like for them when their husbands went off on jaunts for months. Their stories would make a book I would like to read, but it's very unlikely to exist. If the women of today struggle to get their stories heard, then those of times past are even further from reach. There's a gap, a chasm in history.

Of course I knew this already, but I feel the loss of what should be there very personally today.

Then, as I go back to my desk, I remember another book containing a social history of our village. There is something in there I half-read before and I fan the pages between my thumb and forefinger to find it. Yes, here it is. Here she is. Joane Newman: a seventeenth-century 'thrice-married farmer's wife' and I really like the sound of her. I read the single page written and researched by another woman, a contemporary of mine, who has pieced Joane together from the church records and archives. Born as the English Civil War began, Joane lived through the execution of Charles I, the brief Commonwealth that followed and then it was back to the monarchy in a big way. Her life spanned the reigns of a further six kings and queens, from the time before our oak was an acorn through to it becoming a strong, young tree.

By nineteen she had a son by a married man and, according to court proceedings, a physical fight with his wife. She lived like the times she was in: tumultuously, outrageously even, with the key players changing again and again but managing, against the odds, to remain the lead in her story. One farmer husband, two farmer husbands (they kept on dying), four sons and a daughter later and, at forty-eight, she married for a third time: another farmer. She outlived him too and, other than her death, the last mark she left on paper was her ownership of a large piece of land on our road – near, or perhaps even on, this very spot. It is not much – not five hundred pages of walking and musing – but still Joane has forced her way through a gap with the help of a woman three hundred years younger.

I close my eyes and try to see her. There's nothing on this page about her hair or whether she walked with a limp. I'm aware that she is all her ages at once: baby, child, woman. But when she finally settles behind my eyes she is seventy or so, done with husbands, ruling her own roost. Joane would have kept hens on her land, knowing exactly what they needed, when to put them in with the cockerel, how to kill and pluck the birds for the table, how to keep the broody safe from the fox. She would had been doing it her whole life: small-time poultry-keeping to feed her family and put a little extra in her purse.

This, I realise, might be why I can't find anything about the development of egg candling. As a rule it has been men that ploughed, made machines, made agricultural policy, surveyed what they happened to see and made it truth as words in a book. A little coop, a hen, a hand and a candle won't be found in the histories. This knowledge is women's knowledge: passed from grandmother to mother to daughter until it wasn't needed and the chain was broken. Along with kitchen gardens, the herbs of birth and death and all the struggles and joys they existed in and around, this is one of the quiet backstories that no man bothered to write down and few women ever had the chance to tell.

If I want to know what it was like to hold an egg to a candle, what clusters of hurt, love and hunger had led to that moment, I will have to find another way. Joane Newman was a fact – a person – and she only needs to be re-animated. I open my window, take the outside air into my lungs and let it out into her flat outline made of dusty records. Nothing. I try again and then a third time and, finally, her shape begins to swell into three dimensions: fleshy, palpable, buoyant. This work is not the same as reading a hard-edged document and making bullet-point notes, but maybe it is better. This is a record of the past whose chest expands and contracts. It – she – has a smell; her skin tastes of salt, bacteria and linen and as she moves – a creation made of her and of me – the curtains flap at the disturbance. Joane and I are meeting on earth that knows both our weights. We look, we wait and then she speaks.

*The year has turned towards another summer and I'm outside in the dark again, without a lamp of course. No matter, it must be done and I like to do it – to see what others don't. Here biddy, biddy, biddy, don't you peck me, broody as you are. It's only my hand, the one that gives you the corn. Good chick. She is a good biddy this one, six summers of sitting on a nest for me and when she's done with the one clutch she goes to the next. And here they are, on their bed of hay and feather – eggs as warm as the fireside.*

*I'll steal half of them – that will do – as it takes more than it did before to make my back bend this way. I'm older now – three husbands buried older – and there is a gain and a loss in all of that. I only ever really mourned the one: Robert. I think of him still and feel his hand on me. He was not my first, but he was the first to walk me into the church, the only one who looked me in the eyes and kept on looking. He'd already taken a wife, but she was in the churchyard all too soon and he had a girl child still in the cradle. I had one also – a boy – and the memory of the back of a man as he ran home to the wife he'd hidden away with falsehoods. I kept walking though, held the baby* to my breast – weaker vessels, they say – ha! Oh, I'm sure the tongues wagged aplenty again over Robert and I – the two of us making the best of it; me with the baby and no man, him with the baby and no woman. But I had stopped listening to the gossips by then. And it was more than that soon – he let me find my own way across the bed to him, which didn't take long once he'd shown me his smile with just one of his brows raised.

There now, that's enough: enough eggs, enough even-ing reverie about the dead. I am needed by the living for a while longer. I push the door and settle down to see what the biddy has been up to. Take care, Joane. Slowly. I have known myself long enough – too hot and hasty – and it would not do to bring this shell too close to the flame. It takes a while for my eyes to turn this light in the dark into something, but I think I see it. I did tire myself today though – gathering wood betony to dry – and my eyes are not as sure as they used to be. I'll call the moppet in, she must be good for something and it isn't sweeping, that's for certain. Careful, girl! I wait. Never the thing I like to do, always the thing I have to learn. Good. She sees it too, she says: the start of life and the lines trailing from it. Thanks be to God, thanks be to the biddy and thanks be to me – for now, that is. I haven't journeyed through so many winters and failed to learn that nothing is certain. But this is a good sign: the cock is doing what has to be done and there's a hope of the year turning, new biddies and another nest. And I'm sure, then as now, I'll hold the eggs up to the light of candle to spy a little on the future. If He allows me to stay.

I watch Joane put the eggs into her apron – keeping them warm with her body – and a few minutes later, in the distance, she slides them back under the hen who gives a growling squark, but still doesn't peck the woman who feeds her. Then she goes back into the house to live until she's eighty-five– more than most, but not as much as she deserves. She has returned to this place now though and I think we get to carry on our conversation.

The computer screen has gone to sleep and I see myself in it: happily drained by the late night, the hatch, this quest and now by something else – an infusion of shared imagination and memories that aren't mine. Underneath there is still a fast heartbeat and something that could be the moment before panic, but it is spring and I have chicks and I want everything to be fine. I want to be fine. I want to be like Joane: wise, stubborn, brave, tough. That's why I press the keyboard one more time – I want to know more and be certain that this tale she and I have just told is true.

Then, and I'm not even sure of the series of mouse movements that makes it appear, I'm looking at an oil painting. The scene it depicts, the caption tells me, is from seventeenth-century Sweden, but it's not – it is a glimpse of what has just played out for me. The brush strokes work to show an evening kitchen: the lamps are glowing and the sides of the room are dim. There's a young woman in the foreground with a linen cap and apron. She wears a look of anxious incompetence – the moppet (sharp eyes, bad at sweeping). She's not looking at me but deferentially to her left, where another woman sits. This figure is turned away, but I know her face – though only a sliver is visible – and it matches her greying hair: practical, a little wayward, wiry and full. She holds herself with power and contentment, a could-be-unleashed-any-moment energy and a tiny hunch that speaks of new stiffness. This is the body of someone fully present in their life, a life that has been quite the thing. She sits in front of a basket that looks like my own. The woven willow is full of white eggs and she is holding one up to the light of a candle. I zoom in and see the cracks on the oil paint's surface and the curve of the side of her face as the flame illuminates it at the very moment she sees the red veins. The start of life.

Here she is – Joane Newman – and I was right. This painting is enough for me to feel sure. She did candle eggs, women candled eggs and shone their lights on many other hidden things besides. I will make myself responsible for finding more of the gaps and colouring in where their lives should be. When I can't, I will try and enjoy the uncertain blackness, knowing that it is made of layers of thick, dark velvet to wrap around my shoulders as I walk my two acres. I'll find their threads myself with a candle in the dark.

The children and I are visiting the three-day-old chicks who have been successfully adopted by Frisbee – my best broody hen. Once the little creatures were ready to leave the incubator, I slid them under her, using darkness as my cover, and hid in the corner to keep watch. She raised her head in what looked like surprise when she heard a chirp and felt a movement – but was happy to go along with it. And today, both chicks and hen seemed delighted with each other as we top up their food and watch the little ones taking speedy dashes around the cage before pushing themselves out of sight under the hen's breast.

I head back inside with Sofya and Arthur to where my parents are sitting on stools in the kitchen – an Easter-holiday visit. It is bedtime and there's no fuss from the children because my father has a story planned for them: one of the famous 'made-up' stories that my younger brother Liam and I used to beg for. An unstoppable, overflowing milkshake machine has probably flooded the high street with an ensuing river of strawberry, chocolate and vanilla and someone will have to drink a heroic path through it and swim to the off switch before the world drowns in artificially flavoured dairy products. I tidy up a bit, glad to hand over bedtime while Jared makes dinner for the grown-ups. My mother is hard at work too. Washing up speedily, staying busy as ever. As I scoop up 'Catty' the toy cat, 'Doggy' the toy dog and Owly (yes, a toy owl) I notice she has frozen with the tap running, midway through using the scrubbing brush on a green mug. The kitchen light spotlights the copper in her brown hair – still her own colour and only a suggestion of grey.

It is hard to look away, so rarely is she ever still like this. A human whirlwind, a woman of constant motion who has been suddenly stopped. I tell myself that she's thinking about a new plan for her garden or, more likely, an idea for mine. The reassuring voice inside my head chatters in a similar vein about what I am doing. It insists that I am getting on with spring, feeling better, turning what's needed to the light. And some of this has the look of truth. But. I know my mother is not thinking about gardens. She's thinking of Liam and, in this very temporary cessation of distractions, she is letting herself feel his not being here and thinking about why.

Now I am thinking about it too. That he and I don't stop milkshake machines together any more, that I did not invite him here for Easter, just as I didn't invite him for Christmas. He's probably alone while we are all together, my father telling the family story to my children, my mother doing our washing-up. We have been using the dimmer switch so that the reality of this doesn't make its way out of the shadows but occasionally, inevitably, it does. It is clear to me, from the angle of her shoulders alone, that she struggles to bear it and I don't blame her – because sometimes, when I forget to distract myself, I struggle to bear it too. Not now though, not in this moment. As I look at my mother's back and the sliver of the side of her face I feel very little, even less than the blankness I have cultivated all month – as if the present gleam of her emotion switches the last of mine off.

I am protecting the dahlias from what I hope is the last threat of frost. As we have got closer to the end of April, the plot's growth has gained momentum. We've started eating from it again: salad, radish and asparagus. We already have enough flowers to fill the vases: welsh poppies, bluebells, tulips, irises, aquilegia and even a rose in bud; the beginnings of abundance.

Our chicks are growing fast: feathering up and revealing something of what they will look like when grown. 'Have you fed Frisbee and the Chicks?' we ask each other, as if they are a girl group from the 1960s. As I weigh down the dahlias' protective fleece with stones, Jared, Sofya and Arthur watch the chicken family take a tour of the garden. The one that Arthur has named, for reasons unclear, 'Shetland Pomp' finds a worm and legs it away from his siblings who give chase. They all get lost and start hollering: high-pitched emergency chirps that make Frisbee spring into action and round them up in moments. She's a good hen. A good biddy.

An hour later, as I finish up outside, she's waiting sensibly by her house. The chicks are all beneath her and one blond head peeks out from under each wing. She is so in tune with their every step and noise; balancing their need to learn and explore with protection. She makes her body into their home – a bed, a hot-water bottle any time they need it. I open the door to the coop and the chicks stream in after her as if they were one creature. Part of me wants to shrink and join them, burrow underneath and stay there, warm and simple. The numbness is lifting, the springtime distractions are going over and I can feel the destructiveness of March still inside me, where it has been growing, a fruiting body of dry rot in the cellar of a seemingly well-kept house.

I worry that I won't be able to keep it at bay much longer. Whatever happens to me, by the time the leaves finish unfurling on the trees that overhang this little wooden henhouse, the feathered family will have been dissolved. Somewhere around eight weeks after hatching Frisbee will decide that she's had enough. She will switch in an instant; her maternal noises and welcoming wings turning to vicious pecks and chases. At first the teenage chicks won't be able to believe it and will keep trying to cuddle back into her. It won't take that many stabs of her sharp beak to drive the message home. The bond will be broken swiftly and brutally and those early-spring nights warmed by each other's bodies forgotten, a convenient amnesia in their place.

PART TWO

SUMMER 2019

# MISMATCH

It is finally warm and dry enough for us to walk the children to school across the fields. We tell them the plan and Arthur's face becomes a little pointy-chinned circle of anger and misery. He tenses his body and then releases it repeatedly in a jerky explanation of how unreasonable we are being. He feels things deeply, this one. His husky laugh is generous and frequent, his concentration and interest in things deep and wide and he matches this with regular quick flips to furious – a place he has the stamina to stay in for quite some time. There was an eighteen-month period when he was angry about breakfast every day. I thought those start-of-the-day rages would never end, but they did of course; in that slow way children have of not letting you notice or mark another something that has gone forever. Every gain in parenting seems to come with a loss that takes a while to understand.

Eventually the four of us leave the house together and as the children open the gate, any thoughts of hating walking vanish. They run to the top of the field with the goats leaping alongside, thrilled to see their playmates so early in the day. It's the kind of morning that puts a daisy-and-buttercup filter on everything. I pull on Jared's arm when he's midway through scrambling over the fence and kiss him; loving him for making the forty-minute round trip a priority. I am very into the idea, but it often feels like too much of a bite out of my day. I watch the three of them head off down the hill and see them stop after the first stile to watch a startled heron desert his fishing spot and fly off over the woodland of Devil's Hole.

When they are out of sight, I head back to the garden. The old roses – a collection of cerise Rosa rugosa – are in flower near the kitchen doors. They've finally recovered from the inexpert prune I gave them during our first winter and their scent is something else, like the Turkish delight of my seven-year-old imagination. Though I should be inside getting ready for a phone call, I can't help pausing by the flower bed that I've been working hardest on. I've planted groundcover plants such as lamium to fill in the gaps and please the bees, removed a manky-looking rhododendron (which would have killed the goats with a single bite) and added things I love: salmon-coloured lupins, white cosmos and ranunculus. It is a wonderful wait for them to fill out and bud.

As I do a whistle-stop tour of the garden this morning, I see there are two varieties of lettuce ready to harvest and that tiny pears are forming on the trees we planted late last autumn. This is the first year that I've managed to work on enough areas of the plot to feel like I am beginning to get to know it. I have noticed that the ground by the front door is particularly dry, but a couple of metres away the slight slope means it's damper. There's a spot in the east-facing bed where the wind whips over a wall thwacking the leaves and stems and I realise that this is why nothing does as well there.

Slowing down enough to notice these things is the goal I know I should be aiming for. Yet it is not easy for me to hold on to it when everything else crowds in. This is why I walk the plot and set it in my mind every morning and night, trying to feel a part of this place – a kind of a paradise.

Once a week we've been dropping the children with friends and going to marriage-counselling sessions to try to understand why we have moved so far apart and, we hope, find a way back before it is too late. We are sitting on the familiar sofa again now, where it has quickly become clear that our smallholding, the way of life I thought we'd agreed, has become a problem. This land, whether we need it and how we should live on it, is now contentious.

Jared and I have told the counsellor about many things: our childhoods, the best times, the difficult things we've faced together and apart. I have got used to talking like this because I was seeing a therapist of my own until it became more urgent to see one together. My first trip to her office came a year after our move here. After the first intense spring and summer on the plot, autumn had brought a slight let-up and I didn't have more than a vague inkling that there were plenty of jobs to be getting on with, even in cooler weather. Despite the reduced workload I still felt very odd and anxious into winter and, one January evening in 2018, I found myself telling my good friend Clare that I was absolutely fine, but couldn't remember when I'd last felt happy or proud or looked forward to something. Everything was a slog – even arranging to see her for dinner had seemed like a chore. I told her that every little thing took something from me and left me lessened. But she shouldn't worry, I had followed up quickly, seeing her face. I wasn't stressed or sad or anything like that, I just felt wrung out and tired. A dishrag that needs a wash. I laughed as I said it.

Five years previously I'd knelt alongside Clare as she birthed her 11lb 4oz baby on the floor of her living room. For most of her labour I'd sat on the stairs, tucked out of the way, but present nonetheless – a security blanket – while her then-husband and midwife provided all the support she needed. She was doing beautifully, all was going well and I would have been only another body in the way. Then there was a slight shift in her breaths and though nothing had changed, I knew everything had altered for her. She was thinking back to her first baby's difficult birth and beginning to panic so I went quietly in to her, crouched down and said, 'This is nothing like before. You are doing this. He is not stuck. This is normal and everything is going fine.' Sometimes we need someone to notice our distress, meet us where we are in our heads and not wait to be told – refuse to be fobbed off. I met Clare where she was in her panic that day and three minutes later she was holding her son. Five years on, over dinner, she met me where I really was and said, 'Do you think you might be unwell?' She asked it at first and then it became a statement: 'I think you might be unwell.'

I knew it was true but not how long had this been building. Since the summer; since the move? Since before the move? Forever? The next morning, in pyjamas and through tears, I tried to deal with it. I saw the doctor and filled in the mental health questionnaires that were emailed to me. I thought about the prescriptions offered and said thanks but no thanks for now because I wanted to understand what was wrong and why before attempting to fix it. I was signed off work for a while, so

I downloaded meditation apps and joined an exercise class. And when I realised that the four half-an-hour phone-counselling sessions I was being offered wouldn't cut it, I found the nice therapist and used money we couldn't spare to see her every week. I took an unpaid sabbatical and then left the job I had spent the past five years making. People I loved had to listen to things they didn't want to hear, that I didn't want to say, but had to be said. I asked for help against all my instincts and asked again when I didn't get it. Through gritted teeth I repeated the line that I had anxiety and depression, even though it didn't really feel like the long and short of it.

So today I tell our shared counsellor, while Jared listens, that 2018 became the year I was in recovery from 2017. I had tried to take it easy: licking my wounds, making smaller plans for the future so that I could avoid whatever burnout, whatever edge of breakdown I'd been teetering on. It helped a little, the daily panic eased off, and I do feel clearer when I open myself to the plot rather than flinging myself at it. In a roundabout way I attempt to say that perhaps it shouldn't be a surprise that our marriage is shaky if this is my version of self-care. Because it has become clear to me that the closer I press myself to the land, the further Jared feels from me. I need the smallholding, but I also need him.

Five thirty on Friday, the May bank-holiday weekend ahead and the civil servant – no longer a new girl – shuts down her computer and leaves the office. The Tube is hot and claustrophobic tonight, so she focuses on where she is going, her own even smaller holding: a south-east-London, south-facing balcony. She keeps her shoes on when she steps into the flat, crosses the room and then out into the evening to check on her pots. Geums the colour of sweets, purple anemones, ranunculus opening baby pink, and the first orange of calendula. Her young tomato plants are also doing well thanks to the microclimate and she will share their fruits as she shares the flowers: picking them early before work – a bunch for her coffee table and another for a friend.

Tonight she waters the pots in a daily ritual that reminds her to drop the stresses of the day over the edge of the railings to collect in the morning to the sound of the foxes ransacking the bins. It's not straightforward though, the way this place calms her. The plants bind her hands tightly to the balustrade and force her to return regularly and forgo spontaneity. She doesn't know if she owns this place or if it owns her. As she snaps the growing tip off the cosmos, encouraging it to grow bushy and floriferous, she thinks about everything she has had to give up, a lot that has given up on her. Sometimes this green space she's made in the city feels like a consolation prize. Yet it is hers, it is everything, it is all she has.

At home she works on the balcony, in the office she works on the smallholding report, but she blurs the lines too – unable to resist bringing work home, because she is on her own mission to understand why things turn out the way they do. She can't leave the pages she's writing at her desk alone, without hunting for their context. The civil servant wants to understand why this struggle for a square of earth calls to her. To do that she needs to know why the hell they have bothered, and why they bother still – these councils, governments, campaigners

– with schemes and plans to promote something as niche as smallholding.

In this homework a confusing pattern has emerged. It's no secret that holdings have got larger and larger over hundreds of years: market gardens becoming farms, private landowners enclosing as much as they could, common land disappearing and, eventually, industrial farming dominating on the hugest of scales. Yet there have always been persistent attempts to do the very opposite and to preserve the right to have a tiny place to grow what you need. She traces it back to bills of parliament in the time of Elizabeth I and, in a notebook bought for this purpose alone, she lists the many legal fences that have been put up to protect a way of life that was in direct contradiction to the direction progress was taking. A little land was offered by rulers – a token to keep the peasant classes from starving and rioting. A small acreage to hold them in their place. Years passed and it was offered again to the workers, this time to help them realise their rights and rise up. Land to force revolution or halt change and even return us to the purity of the pastoral idyll from which we have supposedly strayed.

She finds another trend too, the good life calling after tragedy, smallholdings in step with cultural and political revolutions, standing side by side with recessions and fear. This comes as no surprise to the civil servant – a woman who has faced many difficulties of late, who has turned to the soil for solace. When bad things happen; when the harvests fail and there's no work, no money to buy food and no land to grow it on either; when there's a war – cold or real – this is when the land's steady call is heard.

From her dives into the library, she knows there have been few successes, that schemes have come and gone, starting and failing within months. The city folk who arrived with their ideals often left after only a few months with baling twine holding up their now-loose waistbands. But she is sure there must be a reason – there is always a reason – why at least the idea of the small life has been safeguarded while everything around it got bigger. A smallholding: so out of step with reality, yet there must be a hidden purpose, after all it is what she wants and she is making her own plans. Or trying to.

My alarm goes off just before 2 p.m. I stop my tapping on the keyboard, pull on my outdoor clothes and head to the shed where I'm confronted by a mess of pots, canes, tools and bottles of organic seaweed feed. Mess like this does something to my brain that I find hard to explain. An alarm goes off alerting me to the mounting chaos. From one side, planning processes zoom in and tell me that I need to sort it out now, or if not now then soon, but when? From the other side, everything else I need to do today, tomorrow, this week, this year, jostles for time and a place in the list of priorities, until my internal viewfinder is a mess of tasks and shame that I can't get a grip on.

There are three ways that this can go. On a good day I can distract my way out of it and on a particularly bad one I might flip to become a frantic, ranting thing. Today it's the third option: my brain shuts off. I'm in the shed but suddenly I can't remember why. My processing speed slows right down and there is a nothingness instead. It is very quiet, too quiet, and I stand there for a moment before muttering, as I often do only half-jokingly, 'What am I doing? What am I looking for? Who am I? What's my name?' The muttering aids a recalibration and I remember that I'm in here for a roll of glazing rubber.

I've come outside now because of the new schedule Jared and I are a month into using to plan out our week. In an attempt to turn down the volume of tasks that are clearly contributing to our discord, we have agreed to dedicate an hour a day to the small-holding and household jobs and find longer stretches to get stuck into the bigger stuff: pressure-hosing the chicken coops, planting out wild flower plug plants, mulching beds and fixing fences. I wonder if it's ridiculous to live our lives by a weekly written schedule – a sign that I am an uptight control freak – or whether we need to be this organised when we have so much to do.

This afternoon we have made time to deal with the greenhouse – currently a partially glazed renovation job for which we are woefully underqualified but that I couldn't pass up when I saw it about to be junked as a neighbouring house was demolished. Jared was enthusiastic; checking with the owners, arranging for our neighbour Harry to pick it up with his trailer and promising to make the fixing-up his project so I can get on and grow things in it.

It's gone 2 p.m. and I am trying to stuff the black rubber in the stupid grooves on the greenhouse panes. This fiddly work is not my forte but I am still waiting for Jared to join me. As one of the ideas behind this schedule is to stop me feeling like the asker, the nagging interrupter, the person who feels guilty about needing something I am resisting the urge to go up to his office and remind him. Since our grand plan was put into action Jared keeps forgetting to look at it with almost deliberate-seeming regularity and so this solution has already slid over into the problem pile. Ten minutes later, to avoid having to take my muddy boots off, I send him a text and five minutes after – more than halfway through the hour we have – he appears at the door.

I try and fail to hide the wobbly sadness I feel about this stupid greenhouse project and our stupid schedule because I know how this is going to go and neither of us is going to acquit ourselves well. I will end up crying and shouting and hating myself for crying about a greenhouse (it's not about a greenhouse) and then try to explain what happened calmly in our next counselling session, while wanting to scratch all my skin off.

It starts to unfold just as I thought. I know I should shut up, but I don't. I do most of the talking and Jared gets more dismissive, then silent and I get more desperate, insistent and unreasonable until eventually I am crying in the bathroom while he goes to get the children from school. I sit down on the toilet seat and, as I do, I have an odd sensation, one that I have been feeling too often of late. It's like putting my foot on where my brain has told me there will be a solid stone step, but because the steps are uneven, I meet air instead. Five centimetres of difference that unbalances me entirely because it exposes the gap between reality and the world I've constructed, and because I'm increasingly uncertain which is which.

I should be sitting on the toilet's avocado-green top, but I can't work out if I've actually connected with it or if I'm still moving through the air. My feet don't feel solid against the lino either, as though I've misread its level too. I don't know whether I am moving or stationary, but I keep crying in an automatic response to what feels like being let down and left alone but which is being presented as me being unreasonable. I flick my head to try to get rid of the increasingly familiar thoughts of jamming sharp things in between the bones of the back of my hand.

These disturbing visions have increased since we started the couples therapy, triggered as I try, and fail, to explain a need that feels obvious. In our weekly slots I've tried to tease out and trace the reasons why the schedule feels important, and how all the promises and subsequent forgetting are making me feel as if I don't exist. Talking in these sessions, spilling out the most fragile parts of me, is not helping. It feels as though I'm pulling out every one of my hairs and each of my nails, laying them on the mug-ringed table in front of us, and then watching the air we disturb as we walk out of the room blow them through the open window.

I am feeling worse than I ever have. Noise is too much: chatter, the kids arguing in the background, radio, the dishwasher being emptied, the cats miaowing, the goats bleating hopefully at me as I prune a rose. A dropped knife floods me with adrenalin and one of my children creeping up and shouting 'boo' makes me scream so loudly that they cry. I need gentle birdsong (not the shrill chirps of frightened fledglings), open spaces and no one to need anything from me. I'm trying to toughen up and stop being so ridiculous, but then I park my car, turn the engine off and feel it roll forwards into the one in front. I brace myself for the impact while reaching for the handbrake, which is already on and doing an excellent job of keeping the car's wheels locked and still.

I need this smallholding to be a simple, easy, happy, family affair with a greenhouse that has all its panes. But it is not and this kind of life has never been like that and never will be. The phrase 'the simple life' wasn't coined by anyone who tried to live it.

Eventually I leave the bathroom with fat eyelids and blotchy cheeks. In an attempt to cheer myself I pick up my new purchase – a little green book about Charterville: a mid-nineteenth-century experiment in smallholding dreamed up by the Irish Chartist Feargus O'Connor. The pages tell the story of the workers' protest movement and his great Land Plan to establish factory labourers on small farms. The paperback opens with a portrait of the charismatic leader and with drawings of the perfect little set-ups he created for those who shared his beliefs. A self-sufficient utopia: low-cost quality houses, animal sheds, dairies and three acres apiece. I read of the idyll that disintegrated to failure, full of reneged promises, missed chances and ruin. O'Connor at the end of his life: broken, erratic, soaked in drink and lost in madness. And then I find her – in the handwritten lists of those who uprooted themselves for a chance at a better life: Elizabeth Nicholson. She moves off the paper quickly and is soon breathing Charterville's Oxfordshire air for the last time before heading away hundreds of miles to cross the Scottish border. This woman is not of my plot but we are in the same place. She looks over her shoulder at me, at the cottage where she's leaving her hopes behind and then squeezes her fists tight to keep herself from breaking too.

*I am walking away from it. From my dream – a dream that took me such a long time of thinking, planning and saving to make real. And now I have the parcel I came with tied up again – though everything in it needs mending – and thanks to the Lord my son and my daughter are still with me too. Charterville. It was going to be the answer, wasn't it? Oh, Lizzie. To think there was an answer; to have* believed it when O'Connor wrote to us 'from Paradise'. It was like that at first too: Paradise on three acres. I'd been hearing the death bell for years, standing at the kirk gate watching the coffins – small ones that they said were made of wood, but seemed to have been planed from pieces of my rib cage. Then another – much larger. I couldn't remember getting to the funeral tea where everyone was talking and eating: I had no appetite, no words. There didn't seemed to be any way to get by in that life and so I decided to believe in his Paradise.

It wasn't quite a year ago that I put my bundle down on the dresser – a dresser! – and though I could hear the skirl of wind outside there wasn't a breath of it in here. We had a bookcase, not that I had any books mind, but I put the jug with a chip that Maw gave me on it to remind me of home. I worked this new ground while the wee ones went to the school. As a child I knew hard work; an early start, a hungry belly and rough hands. None of that was a shock, and through summer I fell into bed happily and slept all night with those two little bodies pressed safe into mine.

But then the harvest was bad. All that work for so little. The caterpillars eating the kale and the potatoes coming up green. We didn't have anything to sell – not enough for us to eat – apart from the eggs. Then, one by one, the hens stopped laying, their combs turned pale and I found one rock-stiff, dead. Three more the following week. The whole flock was gone by December and I was a different kind of tired then. Cooking, keeping the fires burning, washing the clothes, sweeping the floors, chopping the wood for the winter store. The bairns helped, but they weren't big *enough to do much of anything and the little money I had left was soon gone. It felt so empty here – only fields, stretching out forever.*

*The year has turned and the others are starting to leave. This has been a mistake and I can't thole it no longer. I'm not going to stay to see if the rumours of rents never mentioned before and evictions are true. This is not Paradise – it is another place entirely. And so I have to be away.*

Elizabeth continues on her journey back to Scotland. I see her in her home city looking out of the window at the old, familiar view of what she knows now is a lamp-lit impression of a dark night. She lays her head on her mother's shoulder, and finally lets herself cry for thirty seconds before folding up her shawl and putting it away. I see her in my garden as well, touching the spongy onion stems to see when they will be ready; taking in my hopes and mingling them with her own.

It wasn't enough for Elizabeth to give everything to this life. At first the sound of defeat was drowned out by her children's excited shrieks and the cock's early-morning crows, but it was present from the very beginning. An unrelenting noise that crescendoed to insist that a couple of acres couldn't compete with a world that had long been making it far too hard to turn a little plot into a life that could support a new future.

I have put Elizabeth and the greenhouse to the back of my thoughts tonight and am letting the words of *The Velveteen Rabbit* tell themselves to the children, concentrating on feeling their heads fit in the hollows under my shoulders. When they are tucked in, I tell Jared, who is cooking something that smells wonderful, that I'm going to shut the animals up for the evening. It is oddly quiet here now. Not a breath of wind, no cars going by and so I can hear it very clearly: a drum beat in my head. The pitter patter of raindrops, fingers tapping on wood, a snare, hands pounding on the lid of a metal bin, the boom of mallets on timpani, a herd of a thousand horses galloping across tarmac. It is hard to ignore the urgent way it thrums into the air. A disaster, a melting point, a failure, a crisis. Voices of the past, a noise from the future, meeting here in the now of a late-May evening to tell me that it is not enough to find myself back on the land, that this was all the prelude, something new is coming and it is coming soon, whether I like it or not.

# I DISSOLVE

There is a lone metre square of brown tiles just outside our bedroom. Until now I have never managed to work out why it exists among the wooden boards and carpet. Its purpose has become very clear today. With the cupboard open, and the other doors shut, this is a very defined space in which to exist. An enclosed cuboid with a cool floor to lie on, curled up on my side, as tight as I can make myself: an anemone corm, an honesty seed, a speck of brown from a spent foxglove's spike.

I don't know how long I've been here or why I'm crying and thinking about implosion. Implosion: the opposite of explosion, when the weight and pressure of matter causes something to collapse into itself or the space it occupied. A brutal crushing that happens from the outside in. I know that the last thing I did before I lay down was to take my last bit of clarity and use it to text Clare and ask if she could come because maybe I need to go to a hospital. I don't know why or what they would do though. I don't know where Jared is right now. I don't know anything . . . but I'm in here on the tiles and I pull the door closer to make the space a little smaller.

The visions of stabbing myself with scissors in the fleshy part of my hand, of hanging myself from the thin metal beams in the garage, are a constant loop of silent film. I have no idea how to switch them off or get up. I don't want to get up. I don't want to do or be or think. My phone buzzes and after a while I look at it. Clare has rung and then texted. She can't come now, but she will tomorrow. She tells me to get Jared or take a taxi to the hospital if I need to. Shall she ring someone, she asks? Her texts are blue rectangles of worry that slide on to the screen again and again and again. I don't know. Maybe she should. Maybe I should. I half try to do something but suddenly the last fight has gone out of me and I need to sleep here on the floor with my knees against the wall and my back pressing into the tumble dryer. I focus on a ball of fluff and cat hair in the corner and at some point I close my eyes.

I am walking in the garden I know best, though I don't see it as a gardener would. I don't spot the plants that need watering or notice where the bittercress is taking over in the herbaceous border. For me this place is a series of paths, hills and hiding places. A playground: full of colour and places to chase, laugh and ride a bike or sit in the shade with a book. This garden is planted with dens and lost kittens and I don't grow anything here but myself – taller each year despite my mother's pleas to stop and agree to be bonsai'd like the little trees we look at in the place my father calls the Satanical Gardens.

I step away from the back of the house onto gravel and then a sloping lawn where (before my mother got to work on it) there was a view of nature's best attempt at taking back what was rightfully hers: the ruins of a formal Victorian garden. I ignore the obvious sunny route across the lawn, down stone steps and under the rose arch and instead walk along the shady laurel-covered path: a place for whispered games and secrets.

The shed I pass has been castles, caves and prisons and is now full of rabbits. The original two were supposed to be female, but somehow they have become 140 little furry bodies that have me supplying half the pet shops in Birmingham and discovering that I don't know how to say I am overwhelmed.

I hear my just-into-adult-size feet crunch along the paths that my mother uncovered like an archaeologist. Not long ago there was a waist-high celebration of spiky, stinging weeds here. But then my mother determinedly put on her gloves and sent my father and the strimmer in to do battle. She followed behind him; levering roots out of walls, hefting stones and remade this place according to the plans in her head and the instructions of the walls and specimen trees that she uncovered.

Now it is ordered and beautiful again here, with neat edges and – in this my favourite part of the garden – a rose arbour. I break into a run skimming the grass as it drops down sharply and ends, forcing me to jump into the darker world of a woodland dell. Down here the ground is damp and carpeted with ivy and wild garlic that flavours every game of hide and seek. The trees are crowded close and I sit on one of the rotting trunks and look up to the steep bank beyond the fence where the old railway line runs. As ever I am rewarded by a view of the little old palm tree. It grows in a curve of rocks and transforms me, even though I'm supposed to be done with this sort of thing by now, into a castaway on a desert island.

I stay here for a while chatting with the palm as I always did in this childhood of a garden – the garden of my childhood – and realising how much I have missed it and the girl who grew up here. The garden is still there, but my parents wrenched themselves away from it nearly fifteen years ago. She regrets it, my mother, and misses it – as do we all. More than the garden though, more than the usual nostalgia for a place and the end of the over-the-top project to tame it, what we all long for is what we left behind. It's buried somewhere – perhaps in the compost heap or in one of the many holes my brother dug and covered with sticks, heffalump pits for his big sister to fall into. Or maybe it's hiding in the earth that my mother worked tirelessly, unreasonably late into summer evenings, forgetting to stop, forgetting to feed us, forgetting to feed herself. I don't know if it's in the leaf mould or under the pampas grass, but somewhere in that garden I think we left the thing that held our family together. We have been searching for it ever since.

I don't remember when I woke or how I got from the floor outside my bedroom to the kitchen, where Jared is looking pale and worried. I still don't really know what has happened over the past days and hours to get me to whatever new low point this turns out to be. This is what I do know: the children are away, I don't want to go to hospital any more, but I do need something. I know that I have been trying to tell Jared that all is not right for days, for months, for years – but I must have said it all wrong. Though he is capable of dealing with this reality in staccato bursts, he can't or won't retain it and maybe I don't blame him, because neither can I.

I remember that we have been rowing and then I have been spinning out in turn. I know that my head is full of self-destructive thoughts and that though I do not want to do these awful things to myself, the more I see them, the more I feel that the moment of unloosening is nearly here, the more I am afraid that I might. I know that I like the thought of a long, quiet pause and right now I cannot see a way to it that isn't tangled up there in the garage beams.

I am aware that I have been volatile; sitting down and making a to-do list, feeling energetic and positive one minute and entirely overwhelmed the next. I know I have been very quick to forget the distress that's always millimetres from the surface, easily distracted by the next idea or task. I think I have sent up distress flares, but that no one noticed or cared. I know that I can't take the spoiling of any good thing I have done or tried to do and that I couldn't bear it when the flower bed I had worked hardest on – with the just-opened, much-anticipated ranunculus bud – was squashed flat by a felled tree.

I remember that at some point, perhaps today, I went outside and started pulling up the plot by its roots – choosing the things I loved, that were hardest to germinate, the plants I had been most looking forward to watching get taller, grow flowers and spread. I pulled up foxgloves, alliums, sweet peas and larkspur. I pulled up the lettuce and carrots and ripped out the dahlias before they could flower. I wrenched the cosmos and cornflowers from the soil and yanked out the pale calendula that were ready to open. Last of all I pulled up a pink lupin, two years' work, the suggestion of colour on its first flowers making me smile every time I had spotted it through the window. I pulled it up anyway and chucked it on the floor and then ripped the green and sweet-scented mess up with my fingers and screamed. I wanted to reach right to the centre of the earth and burn the skin off my fingers. I wanted to bury myself in molten rock at the world's core and feel it consume me.

I am crying again here in the kitchen as I try to piece it together but become more confused. What I feel is fury at something

– many things – and hatred at someone, myself perhaps. These are coming out as a Morse code of thoughts and worries, everything in little disjointed piles and no logical route between them. This place was supposed to give us what we needed, what I needed. It was going to be a thing of purpose, happiness and somehow it is wrenching us apart and pulling my hinges off. It's good, it's bad, I'm good, I'm bad, I need Jared to help me, he is the problem, I am the problem, I hate myself, I want to go home, this is home, where is home, I don't have a home, I am the home. What a fucking mess of a home this is. I need to see something beautiful. I need the ranunculus that was squashed. I need the lupin I just pulled up. I need to destroy something else. I need to stop. Stop.

I am incoherent, but I tell Jared to move the scissors and then somehow he finds out about the horrible thoughts in my head that I have kept from him until now. He rings my father because he doesn't know what to do and neither do I but it just rings out and then goes to voicemail. We sit in silence for a few minutes while my brain skates over everything and lands on nothing until suddenly it does. I know one more thing: I don't want him to ring my father again. I don't want this to be something else my parents have to deal with. I am a grown woman, of sorts, and I have to find a way to solidify the bits of me that are running in liquid across the floor towards the gap under the door and the land drains beyond. I will not be this person for them – they have been through more than enough already.

Two weeks ago, the first of June, and the sky was as blue as the February day that Arthur and I planted seeds that are now good-sized plants. That midsummer day though, the forgot-me-not blue above fitted the season. I visited the oaks first as I almost always do now and put my hand against them to try to take something for myself, even though I knew I should be giving back too. Not finding what I needed, I turned to the west and walked along the paving path, softened by Welsh poppies growing between the cracks and a layer of moss at its edge. I stooped to pull up some unruly grass and was at eye level with the bottom of the apricot foxgloves growing in a narrow bed against the front of the house. Every time I see them, and the others growing where I shook the seeds the previous summer, the thrill makes me determined to sow enough to view them from every window next year.

I turned ninety degrees to the south and walked towards the Crittall-windowed porch that everyone uses instead of the front door. I found the children already outside, giving each other pedicures. Each toenail a different colour – each toe heavily daubed. I let out a shout because I'd seen the first white poppy wriggled loose of its casing. Papaver nudicaule, 'Champagne Bubbles', with waved petal edges, a spindly stalk and a warm, yellow centre. It's felt casing was less decorative but quite a feat of engineering. I spotted it, as it rested mole-brown and discarded nearby. A perfect oval, split down the middle and gradually pushed up and off by the opening flower. I picked it up and slipped it on the end of my little finger, wondering what it would be like to be concertinaed up and sealed within.

I had my senses open to more than flowers, because I felt very odd – jerking between fine and terrible with nothing in between. I needed a story that day, a slip through the hours into another time or place to find a moment of connection or distraction. I carried on walking, around the lawn, which looked pale and spare in places because we'd finally cut the long patches left for months to let the daffodils die back slowly and allow the goodness to return to the bulbs.

I wanted to be elsewhere, but I couldn't find an opening and just kept returning to the second curve in the bed opposite the east side of the house. It's one of the least beautiful and useful parts of the plot: a wild mess of elephant's ears, random azaleas and spikes of loosestrife dotted with wild-looking roses and a hell of a lot of couch grass. I wondered, why this bed? Why was I being drawn to a scrappy present-day place when what I wanted was to escape? I closed my eyes and listened to the loud noises of summer and felt the grass against the up-curve of my instep. Then I opened them to the tangle in front and began to feel it – the magnetic pull of the weeds and disorder that I needed to get down on my knees and tackle one day soon. There was a story here after all: my story. And within it a template that I should always have known was there.

My parents didn't tell me at first and maybe it was wise or maybe the timing was a good excuse not to have to speak the things they couldn't bear to. It was nearly a decade ago and I was raw and swollen with new motherhood. My arms and eyes were very full of a baby. We were making space for her, finding ourselves bumped up the generational ladder overnight; parents becoming grandparents, children becoming parents. It was a jolt, all of this, as it always is for everyone; reconstructing myself as a person who was the same and yet completely different, my only guide the pattern of the past to follow or reject.

As the months went by, overlapping days divided into feeds and too-short sleeps, I noticed something was wrong. I had to look up from the shadows our daughter's eyelashes made on the top of her cheeks to see it. As I jiggled her furiously to the sound of the extractor fan, hoover and hairdryer in a hopeless attempt to get her to sleep, I didn't ever want to put this baby down, no matter how sore my wrists were, but I had to make space for thoughts of someone but her. I had to think about Liam, my four-and-half-years-younger brother. He used to come and put up shelves for us, make elaborate rabbit hutches and teach my friends to drive off-road. Everyone thought he was the older one at first: so confident, articulate, charming and skilled.

Now something was very wrong with him, or always had been, and not one thing, lots of them. Some were very hard to accept, others almost impossible to forgive. When my parents told me what was happening I looked back at the photograph of him holding my three-day-old daughter and saw what was obvious, but that I had missed. His puffy face and bloodshot eyes; distant, glazed and despairing underneath the joy of holding his niece. How had I not noticed? How could I help? I felt the sisterly tug: guilt, worry, a desire to make it better, to protect him from harm, but then, something new. He had become a danger that I must protect others from. The image of his face made it very clear that I must try to shield my parents, Jared, myself and, more importantly, this new child from the damage he might do.

To the confusing eddy of becoming a mother I added this new direction of spin. What I thought was the foundation for the little family that I was building had flipped and become something to shelter them from. I had been holding the blueprint for a family upside-down and back to front and now that I had the right orientation it looked nothing like I remembered. I might as well throw it away, strike out on my own and build something stronger, more solid.

I told some of this to the therapist I saw a few times when the children were very little, frightened that I might not be able to keep the lid on whatever was underneath how well I was coping with it all. I told her about the earthquakes in my brother's life and the tremors that rocked my parents and then me in between. How it took more to recover from each one as we moved from the first five years of thinking that this was a bad patch that would soon be over, to it becoming clear that his life and therefore all our lives were going to be like this now. My parents' faces and voices pinched and strained and, though I knew I heard only the half of it, that half was horrifying enough. I told her about the things I had to say and to watch – being the one to break the bad news again and seeing my mother drop to the floor. The therapist listened as I talked about the ride of it all: the up and down and down and down and (relief) up before (of course) down again. The new starts, solemn promises, letters, programmes joined and left, jobs started and so many deceptions and dangers. The terror when Liam was missing, the distress at all he was facing and suffering, the rage when I heard what he had done this time. The sadness I put away when my mother said that all of this had taken the joy from having grandchildren or that she would never be happy until he was happy, this very unhappy man.

One day I told the therapist that I had cried all the way home from my ultrasound scan because we had found out that our second baby was a boy and I realised that I was horrified. What if I made a person who repeated the pattern of the generation above? She rarely gave advice but made an exception then and told me to imagine a deep, dark hole. 'Your brother is at the bottom of it and your parents have gone down there after him. I hope they will all find their way out soon, but if you don't want to go there too then you will need to build a fence. Wake up in the morning and build it in your mind. Imagine the hole and put up the fence between you and it. You can see them, you can talk to them over the fence, you can love them and give them things and hold their hands and comfort them, but with that fence up you won't fall in too.'

So every morning through my pregnancy I closed my eyes and I took out my hammer and nails and put the posts and rails in place. I put the boundary up so very thoroughly that I think I accidentally cut myself off from my feelings too – the good and the bad – because, as the first eighteen months of deep contentment after Arthur's birth began to lift, I started to feel blank. It wasn't long after that I had a new kind of row with Jared; it seemed small until I realised that I had been upset and angry with him about this thing for months – and neither of us had noticed. So upset, in fact, that I picked up the wallpaper scissors and cut off all my long hair in a jagged ear-lobe-length protest. I made it into a funny story later at the hairdresser's to try to distract from the reflection of her wide eyes and my red ones.

I was remembering my mask-face in the salon mirror as a shout of 'Mummy!' jostled me back to early June. The children were keen to show off their pedicures and give me a bunch of flowers. I crouched to be at Arthur's height, trying to concentrate on his purple toenails. I murmured approval but couldn't stop thinking about the boundaries my therapist had advised, the messy flower bed behind my son's back and the pattern of the past I had just realised was never abandoned. I had followed it instead – straight back to my parents: the couple who decided to make their lives from a patch of nearly two acres. To my mother: the woman who ran with relief into her garden and worked tirelessly while her husband looked on in shock. And to myself: a big sister winding along the paths and finding happiness in leaves and clouds. When we moved to this smallholding we weren't striking out towards something new, but attempting to return home. I put up the therapist's advised fence with one hand and with the other I changed our lives in an attempt to tear the division down.

"Mummeeeeeee!" The children were fed up with my glazed expression and wanted me to look at the bouquet they had picked. I took in the roses as I tried to hoik myself out of this realisation. Then I noticed they had picked an armful of lupins and ranunculus as well – far too many and far too early. I had been relying on these blooms and now they wouldn't ever open. I tried not to be upset, not to be cross, not to taint the children's beautiful offering and innocent Saturday-morning joy, but I failed. Hating myself for every consonant and vowel I showed them exactly how awful I felt. In that moment, at the beginning of this month, the final root that had been tethering me began to snap. It didn't take more than a single gust of wind to blow me away to the dark place I find myself today.

I have persuaded Jared not to ring my father again and we are talking about what to do instead. I have calmed down a little, a hollowness starting to replace the awful energy of whatever today has been. We are supposed to go to London for a party tomorrow and stay overnight with friends. We should cancel, there is no way I can go anywhere like this, I can't even think about getting from the kitchen to the bathroom. Cancelling seems impossible too though. We have other friends relying on a weekend of house- and animal-sitting for us as their builders rip their house apart. I can't leave them in the lurch. The party is important to Jared, and though he says it doesn't matter, I know it does and I will feel worse if I take it from him. I don't want to stay here without him or put on a brave hosting face for the family who are coming here. I feel like an invalid, as though my skin is paper thin and that I might fall down the Tube's escalators. Nevertheless, going ends up seeming like the best of all the options. A sort of break, a kind of rest, a weekend away from myself. When I speak to Clare she tells me it is all kinds of crazy and begs me not to go, but I say I think it might be weirdly good for me, good for us and that it feels like the path of least resistance.

The next day we are there, though I don't really have a grip on the hours in between. Jared gives me an early birthday present, a pair of sheepskin slippers, and I wear them around our friends' house which is cool and light and uncomplicated: a holiday from my life. Later I dress for the party in a 1970s dress with fluted sleeves that makes me feel like a butterfly. It hits me in all the places it should and I am a fluted swoosh of a person that, miraculously, it just takes a little make-up and a scrunch of my hair to turn into pretty. On the way through London my senses are working harder than ever and my skin feels prickly with excitement and nerves for the evening ahead. Desire too. In a place other than the plot, I notice the muscles of my husband's upper arms under his suit jacket and I want them around my waist. I feel mad, completely crazy. Like I might try and balance on the handrail of a bridge or take my clothes off and walk naked through Regent's Park reciting poetry. But I quite like it. The most fragile, saddest parts of me have been scooped out and put in a paper bag, on hold until Tuesday after the bank holiday when I have promised Jared and Clare I will see a doctor. It's almost like freedom.

We drink champagne at the party, as if it's the kind of thing we always do, and as if I weren't having some sort of breakdown. I feel light, lovely, and find I have things to say, intelligent questions to ask and a laugh that comes from somewhere in the evening air. After an hour or so, Jared and I are at opposite ends of the garden and I am entertaining a circle of older men, waxing lyrical on the joys of keeping goats. I am enjoying being an oddity: twenty years younger than them, a bumpkin dressed as a princess and I have their rapt attention. One, the self-appointed elder, introduces me to everyone else as 'goat girl' and I smile and twirl under the strings of lights that weave through the trees. Then I see Jared through the bodies – his eyes flicking towards me often. I look at him and see someone I remember. Someone I love. Then he is next to me, his fingers lingering on my back. I notice how he smells and see how the rest of our night will play out because it is written in his eyes and the way his hands find the narrowest part of my waist and press gently, my ribs moving up to make room for him. As he leads me away, instructing me to eat and watching to make sure I do, he tells me that the men I have been holding in my goat-girl court are some of the most famous artists in the country. But I don't care about them because tonight he is everything I remember and all I care about.

After lunch the next day Jared and I walk across the London park to an allotment sale. The people here, almost all a lot older than me, are clearly my people. I buy coriander, peppermint, a courgette plant and some peach-coloured verbena with the twenty pounds that the friend we are staying with has lent me and that she will refuse to let me pay back. As we walk back through the park, Jared and I talk. The plant sale has brought me back to myself a little, last night brought us back to each other and the comforting sound of the stream flowing through the trees has loosened my protective grip on my thoughts. 'Do you still want to do this?' I ask him directly, wanting to know if he has changed his mind about the smallholding life. 'Should we sell the house? Move somewhere smaller? Give up?' He does, he says. He values the space, the freedom and he likes where we lives, he does want to do it, he wants to stay – though maybe not as much as I do. But, he adds, he doesn't think he ever wants anything as much as I want everything.

We talk, without recriminations for once, about feeling as if we are in a trap. We have to work hard to pay for all this, but in working hard we don't have time to make progress on the land. This makes me more anxious and then I need the outdoors all the more to help soothe me and then there isn't enough time to do the work to earn the money. We run through our options: downsizing, borrowing money, one or both of us getting full-time jobs, starting this all over again somewhere cheaper. It feels like a relief to have a conversation in which we agree that there is a problem that isn't one of us and we try to solve it together.

The path takes us past a playground and talk turns to Arthur and Sofya: what they get from the plot, the things they have learned, whether another move would be good or bad for them, whether selling the animals would hurt them as much as I imagine. 'I can't make big decisions right now,' I admit before we turn back into our friends' road. 'Neither can I,' says Jared, hinting for the first time at what I have known for a while, that he doesn't feel so good either. 'We can't move, I couldn't cope. Not now anyway,' I say and so we agree to just hold on. We will keep going, make no big changes, take no life-altering decisions, but dig in and hunker down. It comforts me more than it should to have set a goal that requires only that I stay where I am and do as little as possible.

Before we leave I pack my clothes into the small suitcase. My butterfly dress is inside out on the floor by Jared's side of the bed – a walnut whip of seventies nylon. I pick it up and smooth it out, holding it to my face because it still has last night and its suspension of time in its synthetic fibres. Then I pack it away, along with this version of me, knowing she is in there: fluttering, twirling, drinking champagne and taking her dress off in the moonlight because she wants something and for once thinks she deserves it.

Butterflies: peacock and red admiral, common blue, marbled white, small white and orange tip will be darting around the garden when we get home. My nemesis, the cabbage whites, will be there too, desperate to lay eggs on the brassicas. Every piece of beautiful, bird-feeding, garden-racing butterfly DNA is there in the pesky creature munching through my kale. Everything that will make up the double helixes that colour the butterfly's wing pattern is already held in the caterpillar until the moment it hangs upside-down and spins a cocoon or moults into a chrysalis. Hidden inside, and surely it must be excruciating, it dissolves almost entirely, becoming a lumpy liquid of potential. Then, equally painfully I feel sure, it reconstructs itself from the soup to emerge as something unrecognisable, but also exactly the same. The butterfly hangs down again, in almost the same spot it hung as a caterpillar, and it pumps the liquid leftovers from its once grub-like body into its wings to strengthen them. Only then, when it is quite dry, will the butterfly take its first flight over the ground – ground that it spent the first part of its life crawling over on its belly.

We are home and the children are back. As soon as we got off the train and made our way to the car, the paused part of me switched on again and I knew I wasn't well. I am now waiting to see the GP, barely able to look at anything without it setting off a chain of anxious, interconnected thoughts. When I rang yesterday I was told I would have to wait a week, but the surgery rang back this morning and offered to see me today instead. I asked Jared directly and he admitted it. He went there and told them that they needed to see me now, and so they are. The door opens and the doctor who looked at my cut leg in the spring invites me in. He's about five years older than me with an open and straightforward face and he listens as I stammer out how I'm feeling. I'm agitated, trying to keep still, calm and reasonable in this medicalised place of being still, calm and reasonable, but it is hard and embarrassing and really I want to curl up in the dark square of space under his desk. 'I sometimes wonder if I might be bipolar or if something else is fundamentally wrong with the way my brain works.' I get the words out, feeling sick at the taste of them. 'Sometimes I have all this energy, I'm unstoppable and

I can do all these things at once, really fast and then it keeps going, like a whirlwind and I can't stop it. Then other times I'm so muddled and I can't manage, so confused and angry and upset. I don't understand it.'

'What would you like to happen next?' he asks. I haven't prepared an answer to this open question that feels like a gift but find that I know it anyway. 'I feel like there's something underlying this and I want to know what it is. I don't want to take medication until I know what's wrong.' I wait for his response, knowing that last year the other doctors were very keen for me to take anti-depressants and beta blockers. 'I agree,' he says and suggests a referral to a psychiatrist. This is the thing I want – that I didn't know that I wanted – and I feel relief among the chaos. The doctor writes up a referral and explains that I'll get a call from someone from the secondary mental health team to triage me over the phone and I am grateful to him for treating me like an adult and supporting my decision to try to puzzle it out further.

'And in the meantime?' he asks before I leave. 'What are you going to do to cope?'

The plot, I tell him with an actual smile. I am going to work my plot.

So here I am in the garden at the end of the worst month for me and the best for the plants. I have promised the doctor, Jared and Clare that I will just hold on until I get the appointment and so I live only in ten-minute increments, not letting myself think about the past or the future. I no longer make plans or lists and whenever I feel my mind wandering off down an avenue of potentially endless thought, I slam a portcullis down and stop it. I clear brambles and nettles until my arms are shredded and my back screams. I sow foxglove seeds, but I don't let myself consider where I will plant them. I leave next year alone for now.

I walk around the square beds that are cut into the ground in the veg patch. The French beans are at the very start of their climb, against a backdrop of far too many ox-eye daisies. The manky greenhouse has been made more acceptable by the forest of lupins – white, dark pink and light pink – that presses up against its glass and fights for light with snapdragons, a white dalmatian foxglove and purple Anemone coronaria. Behind them chamomile and calendula circle the peas and the parsnips and, in the bed next door, the cabbages are hearting and the chard has not yet bolted – though it won't be long.

Between the long beds at the back, about three foot up my willow arch, the first sweet peas are flowering. Underneath, the squash and courgettes are doing well in their manured mounds and the sunflowers are starting their stretch up towards the clouds. My dahlias are still only foliage, but they look plump and healthy, growing out as well as up, as if to protect the ground from outsiders' eyes. The currant bushes are already covered in as-yet-unripe fruit and some potato plants are in flower.

The goats are watching me hopefully over the gate, bleating regularly to remind me that they'd like a snack of whatever plant is to hand right now. I walk towards them, give them a stroke and tell them to bugger off before heading back to the house, glad that it's Saturday and that I don't need to hurry the children into their uniforms. This is a new way of working and living. One thing leads to another, one task prompts another. I see an aphid and I look for a ladybird to eat it. Then

I spot a weed and find myself weeding the entire bed, realising as I do that the clematis needs feeding, and moving onto that next. A hen's proud 'I laid an egg song' reminds me to refill the water I noticed was running low that morning. While I do that, I will snip the rambling rose that's overhanging the coop and snagging my arm every time I check the eggbox – but only if I happen to have some secateurs in my pocket.

It is fine for the while, as I wait for this referral to come through, to be just 20 per cent of a person: an automaton who is not allowed to think or feel. I can't really work like this or be a friend or a nice mother, but it is what I can manage and I feel okay as long as I stay in the bubble of each moment. The garden has laughed in the face of the destruction I thought I had wrought upon it and has recovered much faster than me. Most of the plants I stuck back in the soil, full of remorse before we went to London, have worked their severed roots back into it and have new leaves already. It's all beginning to look lush, albeit smaller after the surprise prune.

I have been pruned back hard too, but there is no new growth yet. Perhaps humans are less resilient and there are more layers that need to regrow and fuse. I am not ready to do that yet – I don't know which things I want to let grow back. So I carry on waiting, anticipating the ring of the phone or the flutter of an envelope as it hits the floor. Holding on for someone to help me find an exit strategy from this moment-to-moment living and half-wishing I could stay like this forever, an oddly instinctive way of being, pulling up weeds, sowing the honesty seeds that my mother saved for me and then moving on swiftly to laugh at Harry's black piglets legging it away at top speed, looking for all the world like burnt chipolatas on legs.

# OFFERINGS

The first psychiatric nurse calls one afternoon out of the blue when the house is full of children playing. I have been waiting for the call but it shocks me nonetheless and I wonder how many people are available, with no notice, to have this – the most private of conversations – over the phone. I am awkward as we start talking, not sure how to be. I say, 'Hello', and, 'Fine, thank you. How are you?', holding my mobile as I do on a work call, as I do when speaking to my parents or Clare or booking a visit from the vet. But this is not a light social chat or a meeting with an agenda that I have already read. So I lock myself in the most faraway place I can find – the bathroom – and try not to see that my knees are shaking.

I talk to the nurse for forty-five minutes and he is kind, gentle and human. He calms me quickly, asking what it's like to live in the countryside and mentioning an aunt who used to live nearby. His questions delve into my symptoms, my history, my family, the support I have and what I feel I need now. I'm speaking with a quieter voice than usual to stop phrases such as 'intrusive thoughts of hanging myself', 'almost high, mania?', 'feel like I have disintegrated', 'stopped driving, stopped working, stopped leaving the plot unless I have to' from making it to anyone else's ears, perhaps to keep them from mine. I wonder whether I'm half inventing these things, exaggerating them as some sort of attention-seeking ploy. Then with the same intensity I wonder if I'm vastly underplaying it all.

He draws the conversation to a close and says he agrees with me and my GP; it doesn't sound like a clear-cut case of anxiety and depression and there might be something else going on. It would be a good idea for me to be referred to their psychiatric team. He tells me that he'll be putting the referral through and that it will be a few weeks until I'm offered an appointment, describing where the clinic is and reassuring me that they will do their best to help. He ends the call, lifting one weight from me and putting another with a slightly different shape down in its place. Now two professionals agree that I need to see a psychiatrist and all I can do is carry on waiting to find out what it is that's wrong with me.

For my birthday Jared bought me an annual ticket to Great Dixter – the famous garden designed and made by Christopher Lloyd. Along with the children's gift of seeds and a tool for harvesting apples from high branches, it is the perfect present. I haven't really left the plot since our overnight in London but, feeling a little like a prisoner on day release, today we are at the garden's entrance about to go in for the very first time. We turn into the famous meadow, first planted by Lloyd's mother, and it is loose and beautiful – buzzing with insects, the odd gladiolus providing a hit of magenta and all boundaried by the safe formality of the Lutyens yew hedges. Through an arch a collection of plants and flowers in pots is arranged to look just like a bed. The red poppies with black spots – ladybird poppies, I discover – clash spectacularly with Canterbury bells and eight-foot-tall trunk-like stems that look like giant fennel shrinking us down to a fraction of our usual size. I run from one area to the next delighting in this garden and trying to explain to Jared why it's all so good.

We squeeze along the paths – pushing carefully through the plants encroaching from each side – and pass the old barns where last year's dry stems still hang and swallows perch, looking down at us haughtily. Erigeron daisies grow through the cracks in the stone steps and ferns pop up in shady places. Though every inch is worked with vigour, this garden is not neat. It is soft and full and allowed to romp a little. I am happily lost here in this land of sweetshop phlox and angel's fishing rod stems, which curve daintily to block the way. Jared is smiling, infected by my giddiness and his role in it and I am so glad to be here with him.

In the nursery I give in to the rare treat of buying ready-grown plants – a little basket of hostas and epimediums to brighten up the shady areas of our plot – and then I wrench myself away, knowing that I can and will come back. Jared drives us to nearby Rye for a drink and we sit outside in the sun looking across a cobbled street with a glass of rosé. I feel as if Dixter has loosened a bung in me. I'm excited to get stuck into making our own meadow more of a reality and wondering if we could fashion some of the chestnut hurdles I'd seen at the edges of the famous borders, holding back the most exuberant plants. Then I have an idea for a work project for me and another for Jared. I offer him these ideas and my mounting excitement, explaining how I want to come back here with the children and maybe give them each an area of the plot to work themselves.

We agree that we need to go out more, do this more, sitting close together, drinking wine and talking freely. This is life, isn't it? Not the weird half-existence of the past month where all I am is someone trying to feel nothing. Not the swooping, tense judder of the years before either, but a place of balance – such as we find ourselves in now, the sun warming our forearms as we look at the way the sky weaves itself in and out of the rooftops. A warp and weft of blue and brown. Then I notice my hands. Moving a lot and very, very fast. I am speaking in little touch typist's trills, every word spat out speedily and with few gaps. I feel good: happy, creative, energetic, ambitious, touching Jared's arm with love and having all kinds of interesting thoughts. But that isn't the end of it. There is also a building sense of hysteria, a tightness in my chest, throat and head. I feel my heart rate rise and have an urge to write all this down – all these ideas – to commit them to paper or to make a note on my phone before I lose them. I mustn't lose them.

The world has gone from a very small two-acre patch of breakfast, lunch and dinner, interspersed with digging, to a global wonder of colour, opportunity and plans that ricochet off each other and split into yet more ideas that contain promises, expectations and the great bigness of the future.

'I think it's time to go home,' I say, very suddenly cutting myself off mid-flow and returning to the duller, slower voice of the past weeks. Jared doesn't question me and I wonder, as I put down the wine and whatever internal dam I had lifted for the afternoon, whether he had already noticed that I'd unwisely let go of the hold I have had on myself and on the dangerous energy of the last month. So I get into the car, breezeblock my mind shut and sleep until we are home. As Jared fetches the children from a friend's house, I water my new shade-loving plants and then put them somewhere to be ignored until the appointment that I hope is going to get me out of this.

The letter comes ten days later. The envelope is addressed to me but the 'Dear . . .' inside is a name I don't recognise – an unknown doctor at my GP surgery perhaps. I skim the letter, the black type blurring as I look for something meaningful. After a list of the referral criteria for the secondary mental health service that reads like a reprimand, it says: 'Following our discussions the decision has been made not to accept the referral at this stage as we feel that the patient does not meet the criteria for secondary care services.'

But I have already been successfully referred, haven't I? I spoke to the nurse who said it was necessary and that I would be getting an appointment. He even told me where to park. So I go back to the top and read the page again, slowly this time, to see what I must have missed. There is nothing different in the second read – just the same flat refusal of help – but on the third look I do find something. Following the 'yours sincerely', there is a single text box with a border. It is quite small, but it has 'cc Rebecca Schiller' written inside. Within this box, in this letter that came in an envelope with my name on it, there is almost nothing for me except this: six bald digits of the Samaritans' phone number.

I have been busy shutting myself down for weeks, so I cry only a little and then, after a while, Jared comes in and I hand him the already creased paper. He is angry, confused on my behalf and looks a bit scared. And now I am angry too, furious, and as he is the only one here, I am furious with him. I rage for a while as he tries to help, to offer his arms or some kind of plan, but I shout at him to 'fuck off and leave me alone'. I tell him that I don't want him and don't need him, that he never helps me. Though there are little bits of truth and past experience in this, it is horrible and unfair. I wait for him to go – I would; he usually does – but this time he refuses to, remains calm, doesn't react against me. Eventually I let him put his arms around me and tell him that I am sorry, that I do need him and then I really cry. I stay here with my head in the place above his ribcage for a very long time. I don't mean to reject help, as I have today, as I often do, and I have wondered why it is so hard for me to accept it. But here in my hand I have the answer: this letter with its flat refusal and the offer of nothing but the Samaritans' phone number printed like a dare.

It is evening. I am sitting in front of the unlit fire after a day of feeling rageful and flattened in turn. The heaviness of my body sinks me down into the cushions. I am tired from the extra effort required to stop the sloppy, liquid mess of my being leaking out in sudden gushes and jets. I have to wait, again, this time for an appointment to ask the GP what the hell it is I am supposed to do now. Until then I have to find a way to squash this new upset until it is small enough to tuck away unseen, to be pulled out only when I have become someone who can deal with it.

Any thoughts of tomorrow and yesterday remain out of bounds and so, if as much as a breath of air from my childhood garden or the milky days of new motherhood grazes my cheek, I divert myself instantly. For the same reason the past few weeks have made me shove my dialogue with the plot's history and its women aside. Yet today I miss it, I miss them and I wonder if I have been too hasty in distancing myself from a source of comfort and help. It would be a relief to find a safe way to take a break from the immediacy of every heartbeat. I have to have something to hang on to and there isn't much else left.

I look up, as if a finger under my chin is tipping back my head. Above our log-burner is a mantelpiece and in it a beam with two dates carved: 1989, when the fireplace was renovated, and 1922, the date that the house was built. Pinpoints in time, anchors in history, and crucially not my history. Tentatively I stand up and trace the dates with my index finger and then, decision made, I walk to the old toybox where I've hidden the hardbacks, journals and printouts that made up my scattergun project. Yesterday something was taken away, but today I have been given four numbers: 1922. The beginning of this small-holding, the war recently ended and the world re-forming.

The second psychiatric nurse rings ten days later, just before a friend is supposed to pop in. I am hiding in the bathroom again, talking to a brusque man who, from the start of the call, seems unimpressed. Perhaps he's annoyed about the formal appeal the nice GP made after I showed up and cried all over his tidy consulting room. Or maybe he's always like this. Either way he doesn't listen, anticipates, interrupts and then minimises every-thing I say until I loathe myself a little more and am grateful for his condescension. 'You're a journalist, a writer – yes?' he asks crisply and I am surprised by his tone. 'Yes,' I say, wondering why he makes me feel as if it's a secret and a dirty one at that.

'Yes, I've Googled you,' he comments. 'I hope you don't mind.' I do mind, very much, but I laugh and say, 'Of course not.' Which is what I am supposed to say. He goes on to tell me that he's read my recent work and it seems as though I'm coping and I start to wonder if maybe, because I managed to write 700 words for public view, I am. He interrogates me about whether I was sexually assaulted as a child or whether my husband abuses me now. I say no and he lets the 'no' hang in the air for a while before saying very definitely, 'I can put the referral through again, but it will be rejected. You don't meet the criteria, there's nothing here to refer you for.'

'How about some CBT?' he offers eventually, like when my kids ask for an ice cream and I suggest an apple instead. It is something instead of nothing though and as by now I think I should be grateful to lick the mud off his trainers – giant time-wasting baby that I am – I say yes and goodbye and thank you in an over-the-top way. In the kitchen my friend, who has no idea about any of this, is waiting. I blurt it all out to her and her expressions cycle swiftly through interest and surprise to concern and then wariness as she takes in my red eyes with their purple shadows, the clothes I've creased and stretched by pulling at them and by the pause-snatch of my breaths.

A few days after this call and I am on the hunt for bricks to hold down some weed membrane when I notice that, though it is a little late, there are still enough flowers on the elders to make cordial. I am still giving myself permission to pick things up and put them down at random, finished or not, and so I forget about the bricks and grab Sofya instead. We have a newish tradition of doing this yearly task together, but last summer I missed the small window between elderflowers not-yet-open and elderflowers all-gone-over. This is a spur-of-the-moment making it right. On our way to the boundary that we share with Victor, where the best blooms have been preserved in the shadier spots, we pass the veg garden and I show Sofya my perfect cabbage. It is a globe of pale art-deco green, firm to the touch with smooth waxy leaves, that sits contentedly on the mulch I put down months before. I use a serrated blade to harvest it at the stem and then cut into the centre nervously. This is not my first cabbage rodeo and I've been conned by shiny outer leaves before, only to discover that inside is a slug theme park full of tunnels and slime. This one cuts cleanly, falling open to reveal a firm core and tightly packed layers that promise crunch and a light spice. I cut two wedges, giving one to my daughter, and we eat them raw doing a little 'screw you' dance to the cabbage whites.

We meander on to the task in hand, scrambling through the rubble in front of the hedge. Sofya picks what she can reach and then puts the sickly smelling flower heads that I hand her in a bag. We take them inside and wash them gently, stripping off the leaves and stems and then leave the damp umbels to dry. I need this evening to find bottles, sterilise them and check we have enough sugar, lemon and citric acid, but tomorrow we'll assemble the mixture and infuse it for twenty-four hours before straining through a muslin.

When the blooms are upturned on tea towels I make the rest of the cabbage into a spiced slaw that we eat for lunch, crunching it with the doors open, looking out at the goats, who stand on their hind legs to eat the rambling rose growing through the apple tree. Then, in an uncharacteristically domestic flurry, I make a gooseberry and thyme cake, which Arthur helps me decorate with flowers. Gifts from the plot that I gift in turn to my family.

The cake cools and while it does I read a letter from the local mental health service with the results of an online questionnaire designed to work out how unwell I am. I skim it until I get to the part with the score, which reads:

Your Patient Health Questionnaire (PHQ9) score was 13 indicating moderate symptoms of depression and your Generalised Anxiety Disorder (GAD7) score was 19, indicating severe symptoms of anxiety. Your CORE (Clinical Outcomes in Routine Evaluation) score was 24 indicating moderate to severe levels of psychological distress. You reported some concerns regarding risk to yourself. You stated that you have intrusive thoughts of harming yourself or suicide. You stated that you have no plans to harm yourself and that your husband is supportive and you also receive support from your GP. You agreed should your symptoms worsen you would contact your GP or use the helpline numbers provided. You have been accepted for CBT and a practitioner will be in contact with you when an appointment becomes available.

This letter is no better or worse than expected – except the end doesn't seem to match the beginning. I look up the scores because I want to understand them and see that despite completing this questionnaire a couple of weeks after my lowest point, when I had already blanked out much of my distress, I snuck under the arbitrary line between moderate-to-severe and severe psychological distress by only one point. I feel a tunnel of anger and disappointment open to suck me in. I could be back in the state I was in last month – much worse than twenty-four points – in minutes if I think about this any more. So I put the letter in my desk drawer, where it nestles with the old Post-it notes, notebooks, broken pens, expired debit cards and unopened bills. I go outside and I get my spade.

In the warmth of the afternoon I continue with the only plan I have – attending to the rotting mess of grass clippings, straw, manure and chewed apple cores in the compost heap that needs its sides throwing up on to its top. The additional heat that radiates from the under-layers of decomposing muck makes beads of sweat trickle down the back of my neck, between my breasts and along my spine. Sun, heat, sweat, arm, spade, muck. I stay with this rhythm, in the present until it is too hot to continue to look down.

I lift my red face towards the breeze and see that it is easier this way. Cooler and more straightforward to look up not down. Simpler to see the starlings move in front of the clouds than to look at my boots and be forced to remember myself. A person. A score of twenty-four points. Moderate to severe. Could go either way. No, I won't look down. I will look up and give myself the sky and the birds instead. And I need to give myself something else too. Something found in that date: 1922. Yes. I don't need a name on a birth, marriage or death certificate in this moment. When I call someone from the land, she comes to me with no more than this. Four numbers: 1-9-2-2. I watch as the digits develop a heartbeat and lean tired forearms against the fence. Then, like me, she turns her eyes to the sky.

*He said it was easier to look up and see the owl, a barn owl, hunting for mice and rats. He said it was easier for him to look up than it was to stare at his boots or across at the line and the wire and . . . but he always stops there. Never tells me what was beyond, what it was he was turning away from – he talks about the owls instead. Explaining that they hunted in daylight sometimes, because even birds knew that everything was different and none of the normal laws applied.*

*I broke some rules too back then, though never in the daytime. But I missed him as much as I dared. He didn't write often and when he did I'd rip the envelope open so quickly that I'd be afraid I'd torn the letter too. I'd read in the yard – listening to next door's kids shouting at first and then, later, with an ear out for the cries of our own – my own.*

*July maybe – two years into the war – he wrote that he missed us and that the gulls were diving at the owls now, mad as hell that they dared to come into no man's land at the wrong time of day. There was a bit about the cathedral and the whizz-bangs too – though I had to guess at half of it because of the censor's marks. 'Wallows' apparently building nests in the cathedral, no matter the shelling, the 's' lost under a black pen. Wallows. Wallow. Oh I wanted* to but I couldn't, you had to keep going, you see – everyone said it. Just do your bit and wait for them to come back – maybe in the spring.

It was longer, that final letter – longer than any he'd ever sent – and just before the end he wrote that he didn't understand why they didn't scarper, the swallows. They could, after all, they had wings and if he'd had wings he'd have been off on them in a moment and up in to the clouds.

That was the last I heard until the telegram telling me he was in the hospital. Wounded in action. Action – funny really given that he didn't move again for months. Not a thing wrong with his legs or his back– but his mind wasn't so fast to recover. Shell shock they said, though we kept it quiet because people will talk – though not us it seems and not about that. I see it in his face still, in the way his eyes move and how he holds his mouth – how he holds me. When he was up and about and the war over they finally offered him something, a new life – a plot of land – and he said yes without asking me. He came home with drink in him and told me he'd slammed his hand down on the table and said, 'Yes bloody please and thank you very fucking much', and then signed his name.

Mud. We have a lot of it in this new life and I'm supposed to grow things in it. A market garden, which he's working the way his father showed him, and I'm doing what I can, even if I didn't want this. I didn't want the war either. I didn't want to spend the first three years with our boy on my own biting my nails, getting thin and taking comfort where I shouldn't – making the soft parts of *myself hard. Not wanting these things made no difference to them happening.*

*Today we are picking the seed pods and my little boy is helping because he likes the sound of them dropping into the basin. He's a little way from me, the man who's supposed to be at my side, moving the hoe over the soil so that the turnips will grow.*

*I look up at the birds and down at the weeds dying in the sun, across at my chickens pecking in the field, up at the hedge to find where the wren is nesting – and I think that he is right, it is easier this like this. There's a gap between us – things we haven't said, things we won't say – so we have to find a different way. For now it is better to look up at the sky than down at my rough hands or my husband's face or at my little boy, who might not ever know the man I did. It is better to look up and wait for the day we might all tilt our heads at the same time.*

She's in the corner of the field looking for the wren's nest. Her face – younger than mine – calm but hiding the dark inside. This smallholding was offered to her family as they were to tens of thousands of demobbed soldiers, as if it were the next logical place to go after the trenches. A remedy for trauma – peace connected to war by the mud.

She is fading a little, but not completely, as I fetch my secateurs, make a cut in the dry stem and place the calendula head in a bowl on the picnic table in front of me, closing my eyes. The first plums are ripening enough to give off their scent and the sound of tractor engines doesn't quite cover the chirp of a cricket in the long grass. I keep my eyes closed as I loosen the seeds into the bowl and hear them hitting the china with a musical chink; pennies in a wishing well, rice on the floor, acorns on the paving slabs. I feel the past in the strangely familiar movements I'm making. An obvious purpose – something my fingers were meant to do. Like the moments before my daughter was born and my body expelled her without my brain getting involved. Seeing my son's face for the first time and thinking, 'Oh, it's you.' I am grateful for this knowledge that has been waiting for the right moment to escape from my DNA and want to give something back in return. I open my fingers and then scatter seeds in an arc that spans the summer's zenith before falling in light taps onto the ground.

# DJENDJENKUMAKA

This evening, while putting the goats to bed, I see that the Victoria plum tree is ready for us to begin picking. Nothing tastes as good as these plums warm from the tree. I pull one down, splitting it in half to check for little worms and, finding it clean, pop it in my mouth. I close my jaw around this taste of the harvest season. I haven't brought a basket out, so I stuff the two large pockets of my denim skirt with as many ripe fruits as I can and then use my T-shirt to make a bowl of sorts for the rest, exposing my stomach to the evening air. We'll eat some tonight, put a bowlful in the fridge and Jared will make the rest into batches of jam and sauce stored in precisely sterilised jars, sealed with wax discs and labelled 'August 2019'.

It is not only me that is out harvesting tonight. The sound of farm machinery comes from all directions. Combine harvesters, tractors and trailers stacked with bales of hay drive past with increasing regularity. At this time of year I see the long, hot days our neighbours work – sixteen hours from first light until the darkness falls – and often into the night with lights shining and at race pace to get the crops in before the weather turns. The village butcher, whose flocks run near our field, tells me he hasn't had a holiday in twenty years and that tight margins mean the days of paid help are long gone. He does it almost all himself on top of being in the shop butchering and serving. Victor talks to Jared about the low price of meat making a mockery of the care he puts into his animals.

Our yield will require much less of us and yet I've still managed to neglect crops at key moments and now find myself frustrated by a lone squash on a huge vine and the pepper plant that's only just in flower now that it's too late. The soft fruit, however, is doing well and it is the place on the plot and in the season where Jared and I have finally found harmony. He has become king of the fruit cage and is naturally more suited to the methodical work of picking from the branches that I have pruned, fed and watered to get to this stage. I love seeing him go out every evening, often with the children in tow, to fill bowls with blackcurrants, red currants, raspberries, the beginnings of the damsons and blackberries from the hedges.

I take the plums in to him now and, after finding my basket, start collecting the Discovery apples. They are perfect fruits: white flesh shot through with a blush of pink, and are worth risking tonight's windfall wasp festival to gather. I fill my trug, throwing most of the damaged ones into the field for the delighted menagerie but keeping a couple on this side of the fence for the littlest chicks. They squeeze through the gaps and peck at their treat, happy that the bigger birds can't follow and drive them away. I watch them eat, wondering, as I always do, whether I should heed advice not to feed them the cores and decide as ever that I don't believe that the odd apple seed will do any harm.

From this corner I have a view of the trees: the productive ones of this orchard, a lone ash at the garden's boundary, the half-hidden cherry that surprised me with fruit in our first summer and, dwarfing them all, the oaks. All so different in size, leaf, fruit, bark and in the creatures that consider them home. Yet trees all the same, with roots, trunks, branches, leaves and every one of them once a little seed. This is the kind of bothness – in the world and in me – that calls to be probed and dissected. On cue, my fingers move to my pocket and type 'tree' into my phone's search engine, looking for a way to understand by trying to get back to the start.

Tree: an English term that has its beginning six thousand years ago in the root word, dóru. These four letters come from Proto-Indo-European – PIE – a language recreated and re-imagined by linguists seeking the earliest utterances and finding that, without any written records, they would have to play detective. They looked at everything they knew, pieced it together and then made the lost language up.

I dig deeper into our indistinct linguistic past. Dóru comes from the adjective 'deru', meaning hardy, strong and true. Over time dóru, 'the sturdy one', became drew-o then dreuom, treuwaz, treow and, finally, tree. I worry away at this some more, taking the path from 'tree' back in time again to look for the moments when other languages split from this original stem. In Russian dóru became 'дéрево' (derevo); in Sanskrit द्रु (drú) and in the expressive tones of Persian it morphed into 'دار' (dâr). Swathes of the world's forests described with syllables from a single forgotten word.

My search moves closer to home as the bats start to dart through the air and the glow of my screen attracts mosquitos. I explore the Celtic languages. Their lexicons hunkered down in the woodland shade, shifting less than English and staying closer to the past because of it. Here I find that dóru was coppiced long ago – it grew two closely related definitions and 'tree' might not have been its original meaning. Welsh offers me 'tre' for 'tree' – no surprise there – but it also gives me another descendant of dóru: 'derwen'. Cornish has the similar 'derowen'; Irish has 'dair'; Scottish Gaelic: 'darach'; Breton: 'dervenn' and Manx: 'daragh'. The Celts inherited every one of these words from dóru and they all mean the same thing: 'oak tree'.

These clues laid in language have finally led me to something that satisfies tonight's need for understanding: this knowledge that all trees were oak trees at the start. For me, every sapling in every place and time is, was and now always will be an oak at heart. Perhaps, then, the 335-year-old tree that I can just make out in this blurry dusk really is as significant as I have felt it might be. An everytree linking me to anything I need to discover if I let myself go where it wants me to.

Today I am dedicating time to try to be the patient and present mother I want to be. I am hoping that saving seeds together from our current crop – the perfect bridge between the seasons of now and then – might also be the perfect backdrop for parent–child bonding. We're starting the afternoon with bribery and now our sticky ice-lolly fingers are at work in the garden. Arthur has been putting curls of orange calendula into envelopes and Sofya and I are trying not to giggle at his spelling of their common name, 'Marry Golds'.

This is a new job in my smallholding calendar, one I've never thought to do before. Since filling my first basin last month I have been kicking myself for letting all that free seed go to waste in the past. I've tried to buy organic, to support small companies and choose heritage varieties but all too often I've also bought whatever was cheapest and quickest. Now that I have looked into this more I have made a firm promise to save the seed from our plants to use and share.

I know that this new rule is going to make everything a little harder at a time when I need my life to be easier; slowing me down; a new series of jobs to fit into the existing rush that makes me feel overwhelmed in exactly the way I am trying to avoid. Yet I am taking it again – this, the harder path, the extra work – because it feels like the only way.

As the three of us start cutting poppy pods down, realising too late that without their petals I have no way of telling which variety is which, I can't resist the temptation to try to weave a little of what I have been researching into our morning. With all the zeal of a convert I explain that much of the garden-centre seed is specially bred for great big harvests of super-size fruits. The children look impressed with this, as I have been in the past, but then I get to the punchline. We couldn't do what we are doing today with most of those garden-centre plants because they, and their big-brother farming equivalents, have been cunningly designed to be sterile. There's no hope of a future crop, just a trip back to the till and a need to use more of the pesticides and fertilisers that eventually ruin the ground.

The afternoon slides past and before coming indoors to finish sorting our haul we pick the last of the blackened foxgloves. I've been trying not to deliver an agricultural lecture during our precious time together but I've found it hard not to spill the contents of my thoughts out to the two people I most want to know these things. I've told them about freed seed, not owned by anyone – livings made from it, dinners had from it, bees dipping into the flowers that it germinated. And explained that some farmers have been fined chunks of money just for keeping and reusing seeds that they are allowed only to rent. I keep forgetting my audience and using long words or jargon – 'hybridised', 'mixed farms', 'industrial', 'subsistence', 'climate emergency' – watching their eyes glaze and wondering if they are still pretending to listen only because they are afraid of my volatility. Yet when I pull myself back to them now, the way the late summer hits their faces and lights up the world they know, they smile at me as if I have all the answers. We chatter about the things that matter to each of us. The way Amber the goat looks a bit cross-eyed these days and that Belle is growing a beard. How two add two equals farts, according to Arthur, and at what age Sofya can have a phone (when she's fifty).

We take the poppy stems to the kitchen and sit together, splitting the pods and pouring the contents percussively into bowls, trying to keep the varieties separate. After a while they start arguing – a sign they have reached their limit – and I tell them to run off and play. They are gone half a second later, leaving seeds everywhere and a metal bowl spinning slowly in their wake. I try to clear up but the floor is so covered in little grey-black dots that it's futile. If I spilled enough water in here there would be poppies blooming along the grout lines by next May. I stop sweeping, crouch, lick my finger, press it against the tiles and put it to my mouth, thinking of muffins and the Somme.

Eight days later and Arthur is toasting marshmallows over a fire we've made in the field. I am trying to enjoy our family cookout but I can't shake a back-to-school feeling as the reality of the next season looms. Jared is struggling after a summer of leading the charge and, though he doesn't say, I know he needs me to step up and take on much more, and soon. Going from where I was in June, to now, to where I need to be to do that – in just a couple of weeks, with no outside help – is going to be hard. I am trying to acclimatise as slowly as I can, gingerly reaching out to the wider world again. Earlier today I sent a few emails, a tentative response to a request to speak at an event in November, and made myself compose a long-overdue reply to a friend.

As the flames crackle I check the buckets of water we have to hand, just in case, because it is very dry at the moment. The temperatures have dropped from recent record-breaking levels, but it's still hot and there's hardly been any rain for months, the news telling of a spate of countrywide crop fires as haystacks and bales overheat and combine harvesters malfunction. These fires are nothing to those in the Amazon that Sofya, delighted by the kids' newspaper subscription she's been given, tells me about. According to her concerned words an area the size of the UK is now burning in South America. She brings good news though too, sniggering as she tells me that scientists have discovered a type of seaweed that makes cows fart less, but putting on her serious face to explain why that might be good for the planet. A country, she goes on (Indonesia, I find out later) is banning people from cutting down trees and everyone, including her, is very cross with the president of Brazil.

A little later, with the kids quiet in their beds, I am getting out of the bath thinking of the heat raging in the Amazon and of other forests being lost, that are already lost. I find that I am very cross as well – with the president of Brazil, with many others for all that has led to this moment, and especially with myself for not doing more these past few months. Something has shifted in me today but I don't quite know what. As it is still light-ish, I dry myself and put clothes on instead of pyjamas. I take my anger and frustration, my impotence and the push–pull of my thoughts – don't think about this, you aren't ready, think about this, you have to, ready or not – outside to where they can't bounce off the walls.

I walk to the only place that seems reasonable and logical: under the canopy of the largest oak. Since my springtime crash opened my eyes to it, this tree has become a centre point in my world. I come to it often – because I want to and sometimes because I have to, as if physically pushed towards it. Tonight it looks a little different – larger perhaps – growing into the new role I have cast it in: every-tree. It insists I come to it: here is the answer, this is what you need to do, don't turn away, even when it's hard, especially when it's hard. There has been a need to exist in this strange pause that has protected me these past few months: without any other help or plan in place, it seems logical to stay in this holding pattern now. Except that the effort of doing it is becoming a burden. As the days go by, it's taking more and more to force myself to live in disconnected segments, damming the streams of energy, idea and thought that usually pull me forwards and back – part danger and part what I need to feel alive.

I could choose to build the walls in my head a little higher, but I don't want to – I want to listen to this tree, travel where it takes me and feel free. I have to make a choice tonight and that this is the place to make it. There's enough right here in this small plot to feed me stories from the moment the birds start to sing until my voice gives up at the end of my life, but that would be too easy. So many trees, species, peoples, languages, lives and ways of life have been lost while this one has been growing. So much has been gained too – I have gained so much – and yet the balance is off. I have my tree – this oak – while others do not. If I get to have this, I have to use it wisely. I can choose how far I let my mind spin out from this single point: across mere metres, across this country or further – to every person, every forest, every tree. The breadth of the canopy, the size of the fires in the Amazon and the heat of it all demand that I go further. I give in to the thought that this oak of mine has roots that span continents and millennia, sending messages to and from every inch of the planet's crust. That there is almost nothing I can't learn from it if I am prepared for the consequences.

I put my hand on the roughness of its bark and decide to let myself live in a less confined way, trusting that something to help me do it safely will come soon. Now, then, six thousand years ago, here, there and everywhere: there are no limits tonight. I push the metal pin of this new waymarker into the ground, tying a rope around it and connecting it to my waist as I follow through with what I have resolved to do. The barriers collapse and all bets are off as I climb up the trunk, using the ivy's coil as hand- and footholds until I get up into the leaves.

I find a branch, the strongest I can, and I start to swing from it – not looking down, not looking up this time either, but looking out, across, and asking the oak to show me the things I need to see. I swing back and forth and back and forth, my wrists aching, until I am ready and then – as I feel my grip loosen against the build-up of lactic acid, the swing threatening to become another fall – I turn my decision into action. I close my eyes and swing harder, letting go of the oak's branch and flinging myself across, towards, away and over – choosing to jump rather than fall.

It is not an easy landing. Branches smack against my skull, my back, my stomach as I crash through the forest's upper boundary. I thump down through its layers and onto its floor in the shade of the djendjenkumaka tree. I lie muddled for a moment, feeling my arms and legs (nothing broken), taking in the softness of my tissue (battered and sore) and then I look up at this new dóru and forget my bruises in a breath. I have come to rest under a globe of a tree. The canopy stretches up beyond comprehension and there are frogs living in the nooks and grooves of bark and colourful bracts of bromeliads decorating the nine-foot circle of its trunk. A sloth moves slowly – of course – above my head. She was born, has lived and will die in this single djendjenkumaka – a tree big enough for whole lives to play out in its branches. The sharp cry of the harpy eagle begins to break the general babble down into individual sounds. No matter that she is one of the world's biggest, strongest birds of prey, today she sits gently on her newly hatched chicks, looking at them tenderly and giving death stares to the howler monkeys.

It is really loud here, but then there is a bat colony in the branches above and nearly half the known bird species in the world are busy asking me what the hell I am doing in their forest. It is crowded too – in the underwood alone there are hundreds of species of smaller trees, palms and vines. Here at the tree's base where I have landed, the light barely breaks and I have to imagine, to sense rather than see, the thousands of insect species moving and creeping here in this great rainforest of Suriname.

Djendjenkumaka: the Kapok tree, Latin name: Ceiba pentandra. Djendjenkumaka: the Saamaccan name for a tree that tells the story of one of the six Maroon peoples of Suriname: the Djuka, the Matawai, the Aluku, the Paramaka, the Kwinti and the Saamaka. Tonight the djendjenkumaka's leaves shake in the light wind above my head and the sound evokes the Baule people of Africa. I listen harder – a story of many parts and I will get a glimpse of only a very few of them tonight. A people ripped from Sakasso, in what would be stamped, mapped and redrawn as the Côte d'Ivoire, stolen from the place that was theirs just as it was stolen from them. Djendjenkumaka: I say it out loud because, caught up in its unfamiliar syllables, there is the most brutal of sea voyages where bodies that had been people were thrown from the sides into the waves as if they were rocks. The Baule people were forced here to this new continent – a place that they did not know and did not want to be – for a new reality of enslavement. They worked long days that started and ended with lashes and their blood dripped on to this land of Suriname, another land that was stolen, as they grew sugar cane and made secret plans.

Djendjenkumaka: between its consonants and vowels hide a people who risked the fragments they had left to reclaim their freedom – escaping into the shelter of this rainforest or dying in the attempt. A confusion of sounds, smells, opportunities and dangers waited here for them. Yes, the Baule people knew land, farming and plants – they were people who had been in dialogue with the soil for millennia. But not this land, not the million-year-old forest that they now asked to form a wall around them. It was a green mystery that could lead to survival or annihilation. Djendjenkumaka: a word made by a people with no choice but to take what they knew, interpret what they saw, make sense of every leaf, vine and tree in this place and quickly, and compose their world again.

In this forest, that must have felt like their only option, the Baule people made a new life. They stayed, and by thriving they defied those who had tried to diminish them – attacking their former captors and forcing them to recognise their identity, their humanity and their power in a peace treaty. The Baule people turned themselves into the Saamaka and they carved out a new language as they transformed. Djendjenkumaka: the first half of the word evolving from 'egnien', the Baule name for this species of tree. Because, once their eyes had adjusted to the dark of the forest, they recognised it as a tree that also grew in their home, in Sakasso, in the land they once knew best.

Djendjenkumaka: the first part of the word from 'egnien', yes, but the second from 'kumaka', the Lokono people's name for this tree. The newcomers to the forest heard the syllables of the Lokono's Arawakan language across the humid air and recognised another people whose land and lives were under threat. People who understood what the forest could give and how to live in and with it. So the Baule people put together what they knew, 'egnien', and what they were learning, 'kumaka', and made something of their own: the djendjenkumaka. A name for a tree made from two pasts, two presents becoming one future. · A tree, a dóru, an oak with many stories to tell.

I will come back to this place, but it is time to move out now and so, aching as I do, I scramble up the djendjenkumaka and I start to swing again. Thinking of my tree and of the fires sweeping through the Amazon, I jump once more. I land, this time, at the foot of a white oak: a mitigomizh that grows on what was, or is, the Anishinaabe people's lands. There are very different stories being told here as the wind shakes these leaves across the Minnesota sky. Turkeys, grackles, black bears and gypsy moths; acorns boiling in a pan; bark to soothe insect bites, trunks falling for barrels and ships, a people moving west; land sold, pockets reserved and much that had been certain and connected for all of this oak's centuries, and the centuries before, not being certain any more – not certain at all. More trees falling, even tonight, to make space for potatoes to grow.

I put my hands into this unfamiliar soil and dig down to the familiarity of these tubers, a route back to my plot if I want to return, or I could let them take me in another direction: Ireland, Rwanda, Gran Canaria, Peru, where the Ancón-Supe people dug and planted their seed potato and then freeze-dried the harvest, the bitter night air keeping the roots fresh through the long winter and across the thousands of years they have taken to get here to Minnesota and to my vegetable beds.

While I'm down here with my fingers in the earth, I see that I might follow the oak's roots under the rich soil to the island that the Kalinago people called Liamuiga, where they tangle with those of the acajou tree. More histories here too, bending and flexing over time, growing up and out and shedding leaves and fruits and names. A third-century martyr, a famous explorer, a mapping error and an abbreviation later and the acajou tree of Liamuiga becomes a mahogany tree of St Kitts. Perhaps in ordinary circumstances a change of name might not matter much to a tree; it will still grow up and its leaves will fall whatever we call it. But this name change was bound up in a colonial ransacking that saw many of the mahogany trees stripped from the island. At the exact moment their trunks fell, larger swathes of the ancient Wealden forest were being cleared too; the sound of trunks hitting the ground is almost the same wherever they fall. The wood from the acajou and the wood from the English oak both found their way to the corner of a great country house not far from my plot: a delicate side table, a carved dresser – just pieces of furniture displayed in a beautiful house built of brick and stone.

But I know they are more than that now. Old furniture, old houses, old plants, old trees, English country gardens old and new made of money grown in a space where a trunk once stood, by a people taken and on a land being depleted. The money to buy the ground, carve the wood, build the walls and plant the gardens with seeds taken from far-flung motherlands was hidden inside sugar canes. That wealth was hacked from the ground, under duress, by the Bakongo people, who would have known exactly what it was like to be cut down from the place where they had grown – along the Congo river and under the umbrella of its ancient imbondeiro, the giant tree.

From that imbondeiro, I climb, swing and jump for a final time, landing on the plot, my feet on the hard clay and my back against familiar bark. My hands are cut and there is a constellation of bruises spreading across my skin. I light up the deep dark of night with my phone's torch as I go towards the house. The white light lands on the pink and purple petals of the dahlias as I pass, but I choose to stop and look again and see that they are not just dahlias but cocoxochitl to me now, the tubers grown for food by the Aztec people. I want to face up to the horror and complicity and still find a way to grow and pick the bloom, putting it on my bedside table to watch over us as I fold myself into the complexity of the man next to me.

I go inside, my skin tingling from all the places I have been, feeling tired and elated, enlivened and desolate at all I have learned. Jared is busy making more jam, blackcurrant today, and he's spilled sugar across the kitchen island. Little sparkling grains of quintessentially English summer sweetness. A smug, gingham smallholding, an armful of duckling activity – a frivolity. But much more too. Part of our attempt to lean a little less heavily on the planet, to use what we have and stretch it over the seasons, to fan the fire's flames less, to find a link between now and then.

I help a little by tidying the kitchen around him – loading the dishwasher, wiping the surfaces and sweeping the sugar that has dropped on the floor. Grains falling from the jar like acorns from the tree. Our oak shed its first nuts seven years after Joane Newman stopped candling eggs and rested in the earth instead, on the day that a baby girl was born: Susannah Codd, who grew up knowing our oak in its full, acorn-bearing glory. She lived at Codd Farm, which stands, a working farm today, a minute's walk along our road. She was fifteen the day she walked home over the fields from the village with a basket full of foraged fruit for jam-making. The paths and cart tracks would have been compacted hard that summer as she walked over them with the stain of the first blackberries on her mouth, the simplicity of preserves giving way to the crack of pistols and the louder boom of blunderbusses behind her. She broke into a run, bruising the russets at the bottom of the basket, not stopping until she was inside the gate. The next day at the homestead, she made the jam as the story of these sounds came out – a fight to death outside the village inn; the infamous Hawkhurst smuggling gang trying once again to evade capture; a black river of silk, wool, tobacco and lace flowing from landings at Romney and Rye through the helpful dark of the forest's canopy and into the secret tunnels beneath her feet and now mine.

Pretending not to see her supposedly peaceful, gentle and simple world opening up that week to include violence, smugglers and the sound of horses' hooves galloping fast through the trees, Susannah Codd carried on spooning the hot liquid damsons into vessels. She tasted the mixture and revelled in the sugary stickiness that was still a novelty. Her jam was made from a sudden abundance of sugar cane. Sugar presented as a gift on a silver platter that we took, despite the bound wrists of the Baule and Bakongo peoples holding it.

I carry on sweeping them up tonight – the crystalline sugar grains and the occasional lingering poppy seed. Jared asks me why I keep smiling, frowning and shaking my head, why a moment later he spots tears in my eyes, and I try to explain but I don't really have the words. To someone on the outside a jump looks much the same as a fall. Yet I know that I have made a firm choice tonight: the life of complications is the life for me. And even in the kitchen's comfort I am swinging and jumping from the branches into unknown intricacies and interconnections. My stiff body's new bruises riot across my skin: marbled endpapers in a hardback book.

PART THREE

AUTUMN-WINTER 2019

# MAP-MAKING

The first of September brings a pear from the bare root trees Jared and I planted last year. Pears, we have discovered, don't ripen well for eating when left on the branch. Unlike most fruits they sweeten from the inside out, so when the still-tethered flesh feels perfect to the touch, the inside is mushy and past its best. For a couple of weeks, I have been gently tilting the tree's fruits according to instructions from online gurus who have assured me that this small movement will be enough to separate them from the tree. I have been losing faith fast though – tempted to give one a good yank – but I try one more day of gentleness instead and, as promised, the green pear, with a cloud of red on one side, finally snaps free.

We are still eating from the plot, but the summer has never felt more over. The school uniform is ready for its first outing and, without really meaning for it to be this way, I have a busy month of work ahead. It is nearly time, ready or not, for me to open our gate, walk through it and try to pitch myself back to the world as a proposition rather than a problem.

For now, with the pear deposited on the porch windowsill, I return to the safety of the field to shut up the hens. There is a change in the air – I need a jacket – and this new cool gives me a fizzy feeling in my gut. I am waiting to jump again – bracing myself for it – afraid of the flight and the subsequent bump. Yet there's excitement too; an impatience to be, to live, to do and to succeed. It is hard to know whether I am ready. I feel much better, but this progress might be down to the unreal life I've been living and with a few extra responsibilities piled on top I might be back to where I was in June.

The question of whether we should stay here, or if giving up on the smallholding would relieve the pressure on Jared and me – on us as a unit – is still hanging in the air. But instead of letting this merry-go-round behind my eyes, I stop and take in a small section of the plot centimetre by centimetre. I spot a bee species I don't recognise having an end-of-the-day feast on the field scabious. It has a long antennae, yellow-buff head and thorax and a dark, almost imperceptibly striped abdomen. Though it is a little late to see one it's probably a male long-horned bee which, I read, are becoming increasingly rare, perhaps because as solitary bees their lonely lives mean they have no rabble from which to learn. Unlike the social species in the genus, the long-horned bee is less adaptive to changing environments and the new dangers this brings: they are more easily caught out.

This fellow flies off to his uncertain future and in his wake I see that my only Café au Lait dahlia has finally bloomed. I planted it badly, in an overly windy spot that gets too little water, but it has at last agreed to give me a single, odd-looking flower. As its name suggests, this dahlia should be a pale cream with the tiniest hint of milky coffee. Instead its left side is stained randomly with a vigorous pink, while the right-hand petals are moon-pale to the point of glow. It seems as if all the pigment that should have been distributed evenly through each petal has been concentrated in one half – an uneasy split.

I linger out here as daylight fails, the weird dahlia picked for the vase but going a little floppy in my pocket. Night is coming earlier and earlier as we head towards the autumn equinox. Equi. Nox. The twice-yearly moment when the day and the night are as long as each other, in perfect balance – though I'd never thought about what the word really meant until I began to live for the times of year when there is more light than dark.

A few days later and I have arrived in London. The thought of the drive, the car park, the train, the Tube and the day of being 'on' has scattered me across yesterday and this morning. Yet at least some of me has made it to the capital holding this large bunch of cut flowers – a gift for the host of the course I'm running today. Travelling with plants can be awkward – the glass jar that threatens to spill or smash, the top-heavy bunch that wants to topple, the barrage of elbows that might squash and bruise the petals. I am always glad when I've done it though. Today's is particularly beautiful. There are seven varieties of dahlia: Schipper's Bronze, an orange whose name I have forgotten, violet of Blue Bayou, magenta Isabel with her pom-pom-like appearance, the huge white Snowstorm, a little coral-coloured Mignon and the etherial Dark Butterfly all mingling with spikes of larkspur and the milky froth of cosmos. On the high-speed train the other commuters respond to the bouquet with smiles. Pretty flowers. They don't know that I brought my garden with me as an amulet, a force field to protect me from the sensory assault and to keep my brain where it feels calmest.

Now at King's Cross station I notice others, hurrying for their Tube, flick their eyes to the green stalks and bright petals and see their faces relax as they take in a slightly deeper breath. Here in the artificial light of the underground, as far away from the open skies of my plot as I can get in a morning, the sight of flowers just picked from a country garden seems to restore everyone who spots them.

After twelve years living in London, I know the Tube drill well. I move along the platform away from the entrance, through which hopeful passengers stream relentlessly, and stand where I think the larger, double doors of the carriage will open. As I wait for the Victoria line train to arrive, I can't believe I did a version of this scramble almost daily. The station is scented with bodies, tired air and machine oil and when a train arrives it is impossibly full. The wondering if I can remember how to do this makes me too slow and by the time my body reacts, people have already disembarked and a few new passengers have shoved their way on. There is no room and so I hold back. Two minutes later another train arrives and, though I am ready this time, it's the same jostle and red faces. I have no idea how I am going to get on this train and so I wait, hope and look around.

There is a deadening cloak you have to put on before you wrestle regularly with a busy city's transport network. It's a shawl stitched from numbness that protects against claustrophobia and bad smells; like opioids in labour. You can still feel the pain, but you don't give a shit. It has an out-of-body effect that removes you from the experience of pressing yourself against strangers and makes you impervious to anyone and anything save the tiny gap you think you can insert yourself into and then do. Mind that gap and only that gap.

On my first train journey this morning I read the novelist Penelope Lively's book on gardening, in which she writes about the 'noticing' that all gardeners train themselves to do. Here on this platform I realise how right she is. Since I left the city behind and filled my days with active noticing my already sharp senses have an even finer point. I no longer see those around me as blurry faces and instead my brain tries to notice them, every single one. I have lost my deadening cloak. I don't belong here any more.

The Tube trains pull in and out and I am no closer to boarding. To distract myself from my wants-to-tap foot and how much I hate even the suggestion of being late, I focus in on a man who is also waiting in the little crowd nearby. He looks a little like my stepfather-in-law: bald and smartly (though not chicly) dressed. Studious-looking glasses, suit trousers, the top of a white collar and blue tie emerging – not from a suit jacket – but from a high-tech, lightweight cagoule. I'm sure he would call it a cagoule.

He's very trim, but not thin, in a way that speaks of exercise and competitive outdoorsy-ness. The man is in his late forties, perhaps early fifties, and would look self-possessed and in control were it not for the absent-minded way he is chewing the cable of his ear-bud headphones: high-tech, good for running in, expensive. He chomps on them, his mouth opening and closing slowly, and I catch little glimpses of his tongue as it twirls the grey cable. He is not expecting to be noticed here but I see him nonetheless and I am embarrassed – for him, but more for an imagined me who has forgotten to keep tabs on what I am projecting into the world.

The cringe makes me turn away and a moment later I finally manage to squeeze onto a slightly less crowded train using my back to shield the vase from the crush. As we pull out of the station a woman shuffle-runs onto the platform and trips on her too-big shoes. I tense myself against the tumble but we are in the tunnel's darkness before it comes. The breath of a smart man to my right makes its way to my nose. He drinks but hides it and the weight of his secret hunches me over. I straighten to cast off his load and am drawn to a woman in her twenties sitting across from me. She looks at her phone and fiddles gently with a bead bracelet that spells out – ELODIE – letting thoughts of the night just gone remove her to a softer place.

My eyes track right along the seats to an older woman whose clenched jaw and balled fists direct me to the (easily and instantaneously monikered) Expensive Suit Guy standing in front of her, banging her blue-trousered shins with his leather case. My jaw tightens with her irritation grown to rage after decades of putting up with this shit. But I'm pulled from imagining the perfect revenge by the man to my left who keeps looking at his watch and tugging his shirt collar: a job interview, I think, with a pang of jitters, and we are stopped in a tunnel so he might be late. I might be late, I realise, my heart beating faster.

I can't do this. This noticing is too much, especially here in the middle of the melee without something firm to brace against, so I look down to try to block it out but a multitude of feet – each shoe with its own story – does quite the opposite. I look up instead and there is the Tube map – multicoloured lines and dots neatly spaced across the poster. These routes offer a new, calmer view. I start to trace them saying the names of the stations in my head: Aldgate East, Whitechapel, Stepney Green, Mile End, Bow Road – moving east along a strip of green and pink – places I know, places my friends live still. I don't think of the people, nor the shops and street lamps, but absorb myself fully in steady progress from one circle to the next. This diagram is not an accurate representation of distance or direction. It is only a tool to navigate the complex network easily and find our way from one side of this incredible snarl of a city to the other. This map is reductive and were I to try to use it to find my way above ground, I would fail miserably. For today though, down here in the close, hot depths of the overwhelming underground, with tonnes of London clay above, it is what I need to find my way.

Exactly four months after I lay on the floor and wondered whether I needed to be in a hospital – and ten days after my Tube journey – I am sitting in a room looking at a man and wondering whether he is actually going to help me. The CBT I was promised has started and the first appointment has come not a moment too soon. My tolerance for the real world seems to have lowered each time I have been forced to have a break from it. But I can't stay cloistered on the smallholding forever, the overdraft enlarging and parts of me atrophying. I need this man to help me find the right path.

His name is Ben Cherry. He's younger than I expected, younger than me, I guess. I smooth my clothes as he introduces himself and goes through the 'measures' – the results of the online questionnaire I have to complete in the twenty-four hours before every session, despite the miserable little prison of each question's multiple-choice answers making me want to burn the building down. We talk generally, then more specifically and I try and fail not to read too much into his facial expressions and to concentrate on the plan we make to meet next week and start in earnest. I like him. He treats me like a person – one with a working brain – and puts just enough of himself into the room to remain perfectly professional while making it clear that he is a human too. Ben clearly cares about helping me, and gives his time in a relaxed way – as if he wasn't hemmed in by all kinds of policies, funding restrictions and pressures. He is trying hard and so am I. This is what I have been offered and I take it with both hands and decide to make of it what I can.

Jared is walking into the kitchen – back from today's school drop-off. I pour him some coffee, hearing the buzz of a text message arrive as the black liquid sloshes into the white mug he likes best. Within seconds of reading it my calm turns to fury and worry – the parcel I need has been delayed, again. He tries to reassure me that all will be well: he can go and get the tool I'm waiting for, I can do the job another day or maybe it doesn't need to be done this year. His words make things worse though. Doesn't need to be done – ha! My brain has already extrapolated this issue across seventeen different planes of consequence and so all his suggestions sound like baseless pats on the head. As I explain all the ways this delay is going to sabotage me, I keep seeing the clear, straight lines of the Tube map, longing for the calm it offered instead of the current crackle in my skull.

The intrusive stabbing thoughts have mostly been kept at bay of late, but they have been replaced by seeing myself hit the floor in sweaty repeats of fast-paced, punishing press-ups.

When these filled my head for the first time I got down and tried it – hoping to banish the image by completing the vision. It didn't work.

I zoom along unchecked, moving from departure point to destination in one huge leap, skin pushed back from my face by the energy coming from inside me. I don't need an outside force to feel my vision blur and my body become leaden and light – I can do it myself, taking something small and zipping past all the steps between it and a supernova in half a second. The end point and eventual cruising speed are not the issue. What bursts blood vessels and muddles my mind is the sheer rate of acceleration: it's the dramatic change in velocity that causes damage.

As I try and fail to calm down I see Jared's rigid face and I try to turn his eyes into the circles of stations. Places to stop for a while, to rest, to open the doors and let some air in or disembark: one, two. This helps a little and so I look for the next stop along the line, but it is hard to make out because too much is flailing. So I jump, change lines and walk out of the back door.

It takes a huge effort, this redirection, but I try to decelerate gently by looking at the smallholding and seeing only its shapes, its lines, points and edges. Along with the great oaks that run along the fence between our plot and Victor's farm there are willows, ash, cherry and maple trees. Together they form a straight-ish line of maturity suggesting that this division of land has been a physical reality for some time. The other fence reveals something different: a hedge, some overgrown conifers and a whole load of dubious-looking sheeting propped up by metal posts – a much more recent partitioning. With these thoughts of grids and patterns comes relief. My adrenalin is dissipating: this is working.

I have been accidentally collecting copies of old maps of this area for a while and for the first time I want to look at them and work out the physical evolution of our plot. They are waiting for me as I return to the house, squeezing Jared's arm to let him know I have come down from the ceiling. The first document, a hand-drawn tithe map of the 1840s, is no longer flat and brittle but swiftly becomes three-dimensional under my gaze: fingers on the paper, wobbles in the lines, lives hidden in the ink.

I look through a series of Land Registry plans, at old Ordnance Surveys that were updated periodically and at historians' pencil sketches and their painstakingly catalogued accompanying records. I lay these new shapes and numbers over the land that surrounds me and without leaving my desk I walk across it, watching our newer boundary disappear and erasing the one between Harry and his other neighbour for good measure. A huge trapezium materialises with the number '405' handwritten across the grass: an enlarged field that knits three smaller parcels of land, including our own, together. The pasture that, according to this map, was called Church Hams.

As I watch our home dissolve, a small building assembles itself to house a manger and half-full water trough. I switch direction, advance a hundred years and the field divides once more and three bungalows with pigsties, barns and sheds spring up on neighbouring plots. Over the pen-line fence of our southern boundary I am led along the familiar footpath to the village. It cuts across Victor's land – three acres of grazing that were once named Footway Field.

Turning back, I walk along our road to the west, demolishing houses and inserting open fields and woodland with my gaze. I think back to Hasted and the other men who perambulated the county trying to codify it and whether they used maps to find their way.

My footsteps beat a rhythm against the tithe map's dried pulp as I walk the line of Mabel's View – where I sometimes find myself on a run, cursing Mabel (whoever she was) for the steepness of her relentless incline. At first, in the autumn sunshine of yellowing paper, I find that what I want most is a chart of my own. But as I observe the gates moving into the places someone's pen mandated, I realise that a single, static map won't be enough for me. There is nothing available that lays out all the unknown things ahead. If I knew how to make a map though, to revise it over time and keep it updated according to my purpose, perhaps I would never wander off course. What I need is to find a mapmaker from whom to learn.

It is 1729 and the man stops often as he advances towards me. It is not the first time that the Mapmaker has come to this village. Nearly thirty years ago, newly apprenticed to his uncle Francis, he picked up a heavy pouch of tools and followed the older man to chart a local estate. Now the former apprentice has become Master Hill, a revered professional in his own right. He is returned to these parts at the call of John Bridger, a Canterbury man who wants to record land he has inherited and set his weight down firmly upon it with a topography of future rent payments owed.

The Mapmaker likes to work alone and so no one carries his bag today. He feels it pull at his shoulder as he decides where to draw the field lines, the sun deepening the scores that are beginning to divide his face into a series of irregular shapes. He is used to the looks he gets as an intruder in this village of only sixty dwellings and happy to stare intently at the things its inhabitants no longer see. Accustomed to choosing what to include and what to set aside, he takes his time deciding how best to erase the movement of life from this scene. Later, sitting by the light of his window, the Mapmaker will choose a palette of green for the land, yellow for roads and red for roofs. In the places where ditch, thicket and broken-windowed shack are of no interest to his employer, he will leave the paper blank – the gap acknowledging the distance between what exists and what he has been paid to see. The measurements he takes for those who pay him are acts of simplification – constructions of necessity, lies of a sort. They help others but they also help him to cope with the weight of the world's complexity.

The Mapmaker is also an artist. I know it from the flourishes on the finished document's cartouche and from the way he looks at me here on Mabel's View as we pass with a nod. He sees it all, even if he doesn't add it to Bridger's plan: homesteads and holdings, hens pecking, children playing, people at work in the fields, lovers in the shadow of the hedge and paupers' hollow cheeks as they walk past an abundance of food. These are what he remembers and draws in his own time and with his own inks.

He is above me on the hill now – the Mapmaker, the Artist – but he turns, as do I, and we look at each other. The whole village knows who he is, not much is missed by the gossips and fussocks. But he is not of this place so he shouldn't know my name, Mary Turner, nor that I am just come back from the Churchwardens and Overseers of the Poor of this village. The Mapmaker can't have fathomed that this baby, warm and safe in my shawl, is the bastard of John Smith and that I am shamed and called a hedge whore, a game pullet – turned out of doors and spat at ever since my belly grew big. He has no way of knowing that I throw that shame down and trample it toe to heel with every step. He didn't witness my moment of righteousness today when the wardens told that scrub he must give me one shilling and sixpence a week for his child's bringing up.

The Mapmaker can't know any of my history, yet I feel sure he does and I fear it and welcome it alike. His eye is kind and keen as it takes me in – the determined angle of my chin and the way my feet drag slightly on the clay. He looks at me as if he wishes he could make me a map. One to follow into next year when I will have to put my baby in the ground and beyond, when I find a way to move my mouth into a smile again. I hear it then, metal landing on stone and then he is striding away, almost running from me as I turn towards the object he has let fall. A few steps and it is in my hand: round, made of brass and engraved with marks I can't understand. As I look up from the view it gives of my own face, he turns one more time, too far away for words, and with a wave of his hand lets me know that I am to keep it – this thing that fits in my palm with arrows that spin as I turn. Pointing me towards something, pointing something towards me. A way to map my own course.

I tuck it in the shawl with the baby and carry on my way, not looking back in case he changes his mind, and then turn along Brooke Street past Brooke Farm, alongside Church Elms and stop at the pasture opposite where the cows graze. I pull out my new treasure and hold it up to where I am going – the footway across the further field and then through Devil's Hole. One of its arrows points to my path ahead and to a snake-like symbol at the circle's edge. I draw its shape in the air with my finger: a little curve up and then turn, a curve down and then turn again: 'S'. The smooth movement of these arcs comforts me as I shift the slight weight of my child and set off to where the arrow points.

Mary Turner walks out of my skin and out of my field now and I am back to the two-dimensional, tracking her as she takes the cut of the footpath towards the village and then beyond the point where the scan of the old document meets the edge of my computer screen. She has taken the shortcut, but I think the Mapmaker will reach the village first – propelled by his longer strides and regular meals. He knows how to navigate the land-scape without the compass that Mary is now learning to use.

I need to learn these skills myself – I can and I will. But I want him to come back too. I want him to give his compass to me. I want him to put his arms around my waist, to show me how to make the world simple and then complex again.

I just want him to put his arms around me.

I just want him.

I have wanted him since I looked at the brilliance of his inks, since he faced me on the eighteenth-century road and saw my every atom. But really I have wanted him since before any of this. From the moment I saw that, of all the uncommon first names he could have had – at a time when everyone was called Thomas, Joseph or Richard – the 'J' in Master J Bridger stood for Jared.

When the rain stops for a while on the last Sunday of the month, I nip outside to move hay from the garage and notice that the house's render is covered in daddy-long-legs. Not a small cluster, but hundreds of them everywhere. A crane-fly pilgrimage for which I have no explanation other than the forecast for high winds which has had me battening down the smallholding's hatches. I tie up a few roses while I am out here and spot that the hollyhock in the drive has gone to seed. Each flower that turned into a pod has now opened like a cup and the circular seeds are arranged on their sides in a ring, each slotting perfectly into its matching groove like a slide in a projector. It is a beautiful and practical arrangement that I now destroy in the pursuit of more hollyhocks.

As I broadcast the seeds randomly, I am delighted to see the Black Jack dahlias have finally opened. They have steadfastly refused to bud all summer but tonight their maroon-black flowers contrast as planned with their coral-coloured neighbours that have flowered continuously and extravagantly since early July. These different trajectories have annoyed me – unable to complement each other if they aren't open at the same time – so I am glad that now, radiant if a little late, the two finally overlap. I take the last of the seeds to the driveway beds and as I pick my way I spot a flash of purple. Daisy-like flowers on tall stems – asters – something I've had on my list, but were already here if only I had looked and seen them. Do I notice too much or not enough?

I store the surprise and joy of these flowers and come inside bringing an armful of logs with me. It is the first time we've lit the fire for months and the room starts to smell of burning dust; the residue of the summer giving in to the flames; disappearing up the chimney and entering our lungs – becoming part of out there and in here. There are still days of warm weather and blue skies ahead and later I will go out in the dark to pick raspberries – a peace offering for the Jared of 2019. But it is time to fill the woodstores nonetheless, collect kindling and get ready for the coming season. I'm already thinking past autumn to winter – because, as well as slowing myself down, this life teaches me to always be two steps ahead.

# MISSING PIECES

I'm walking away from my latest CBT session and the rain takes me by surprise. I have not yet remembered what it's like to live in autumn and winter, so I haven't brought a coat with a hood, or an umbrella. A few steps away from the building and, though my head is drawn in like a turtle, I am already drenched. So I try to embrace the wet, walking as slowly as I can bring myself to and looking at the sky, fixing on the drops as they drill down and land on me with a splash. I always feel out of place here in the centre of the town ten miles away. The crowds of people, shops pumping music, the smell of lunchtime and so many stories up in the air ready to fall on me.

Looking at the rain helps tune the cacophony out and get back to the car park. I choose this place to leave the car every time I visit the town because I know its location and understand how it works. On any first visit to a new place I brace myself against potentially getting lost, being late, breaking the as-yet-unknown rules of the tarmac and becoming trapped at the barrier arguing with the operator via a crackling intercom. On subsequent trips I cling to this hard-won familiarity, no matter if there is somewhere closer or cheaper, even though I have no idea why something insignificant feels so fraught in the first place.

I can understand then why the therapist started our session by saying he wanted to give me a provisional diagnosis of generalised anxiety disorder. Ben told me that he's not qualified to draw official conclusions but, as he needs a framework, this diagnosis (that is not a diagnosis) could be a helpful place to start. He pulled up the relevant page on his laptop and asked me to read a description of the condition and I did, recognising the smack of anxiety in the strict rules I have made for myself around something as neutral as parking my car. Yet before the end of the first paragraph I knew it wasn't right.

Yes, I told him, I do worry and overthink but I'm often also opportunistic, adaptable, brave – foolhardy even. If I decide to do something, I push ahead quickly without much thought of consequence getting in the way. Doing the right thing often compels me to act even if it means doing something risky and exposing. The bigger the challenge, the more I am up for it: like upending our entire life to move here based on little more than a whim and the notion that it would all work out okay.

Though I was aware of how much more complicated I was probably making this for Ben, I explained that the anxiety I feel is different. It impacts only on certain areas of my life and exists in the small things: housework, planning for holidays, the way other people view my appearance, the house, the washing pile, getting behind on picking up the leaves or updating my computer software. I told the confused-looking therapist that this anxiety is not generalised – it is very specific, though what it is specific to eludes me.

A little over a week later I am outside giving myself a pep talk and searching for signs of next year in the ground. There are mushrooms, toadstools – so much fungi all of a sudden and, though I'm far too much of a wimp to fry some up, I wonder whether they are edible and take pictures of the ones that look as if a fairy could be along any moment to lean on them. There are so many foxglove seedlings growing now – some self-sown and others helped along by me. These little reminders of next summer make me smile despite the sick soup of worry that has been creeping up my body since I arrived home from a work trip last night. Somehow I'd forgotten this all-consuming sensation of being physically ill with unspecific agitation.

I finish my walk and return to my desk still feeling a bit hopeless in spite of all this new growth. Some of the reasons and triggers are obvious. I have CBT again tomorrow. We have a money issue and I am earning what we have by coming up with projects that might overstretch me. Jared is not doing well and his internal state was mirrored and magnified as I made my way through the house yesterday after a week of absence. The resolutions of connection and positivity I'd made on my journey home started to slip away at the sight of the mess, the towering washing pile and the feeling of cold air indoors because the portable radiators had not been brought in from the garage. I tried to ignore these things and not stack them one on top of the other as something else to deal with and to instead be kind – to myself and to my husband.

Then I saw every vase I'd filled with flowers before setting off had turned to a stinking mess of sludge. He knows that coming home to mess jangles me, throws me off and so this felt like a trap or a test. I couldn't hide it and then that made me more certain that I'm a selfish person, an unsupportive wife, and the nasty voice in my head started up again, jostling elbows with a part of me that felt sad and tired and let down.

Since that first summer here that broke me, I know I have been pulling in two directions trying to work out an alternative mode of being. Needing to step up, commit, be more; needing to step back, let go, slow down. And yet, this imperative to relax, look after and be kind to myself comes with a matching instruction to ask for help. I worry that if I let go and live outside those tight little lines, everything will collapse. Last night I could see this reflected in the vases' murky water. It might be just as I feared; if I put my burden down, no one else will pick up the important pieces of it.

Fridays have become my Ben Cherry day so I am here again in this strange mixed-use office reception – with people waiting for job interviews and others, like me, trying to look as though we are here about the administrator vacancy instead of a state-funded mental health intervention. Inside the room the therapist tells me he has been talking to his supervisor and they have come up with a new provisional diagnosis. I'm both pleased to be a special case and horrified to be a hassle as he sets out a different model: 'clinical perfectionism'. I sit up a little straighter – I'd rather be a perfectionist, even a crazy one, than someone with a generalised worry about being alive, though I'm pretty sure this in itself reveals something terrible about me. Ben explains what 'clinical perfectionism' means: a person who sets impossible standards, gets little pleasure from reaching a goal because they assume achievement means the aim was too modest. He describes abject feelings of failure and then complete demotivation. There's a refusal, apparently, to do things you aren't naturally good at and a punishing cycle of working harder and harder, setting tougher targets: all the feelings of failure without the balancers of success.

Something of this feels right, though not perfect – and I tell him that I absolutely see the irony in that. Yet for the first time, I entertain the possibility that I won't always have to carry on feeling like fifty-seven square pegs trying to fit in one dainty, round hole.

At home after my appointment, I try to put my new mechanisms in place: go outside, come back to the internet, find a story, meet the past, learn something, feel better – but it feels like a hollow series of things that I dress up as meaning to distract from how messed up everything is. I am busy hating myself via an attempt to engage with the 1939 Register, a record of who was living where in England and Wales that year. For a while I have wanted to find a tangible person who lived on this plot around the time the boundaries of today were created. In the less bright light of autumn – though it seems a little disloyal – I feel the need to know specific details of a real person. I hope that someone who I can be sure touched these walls in the past will anchor me more firmly than shadows and whispers. I planned to turn to the censuses for this purpose but have discovered that there are no censuses available for the first thirty years of the smallholding's existence. The record of 1921 just pre-dated our house and that of 1931 was destroyed in a fire. By 1941, the Second World War had put paid to that decade's efforts and so, the 1939 Register – taken at the outbreak of fighting as part of the war effort – is one of the few ways I can find out who lived here and puzzle out when and why this plot became what it is today.

I am trying to focus my attention solely on this task but my session with Ben has awakened a swirling upset that's rising in me. Perfect, sneers my head voice, don't be stupid, you are barely competent. Perfect? A perfect bitch more like, can't even be kind to her husband when he's depressed. Selfish cow, stupid weak bitch. My version of putting my fingers in my ears to this is reading the list of properties on our road from beginning to end. I review them again and again but I can't find our plot listed where it should be, however hard I look. The acerbic tones of my inner voice get louder with every failed attempt so eventually I give up and try another research trail in the hope of drowning it out. I follow the records away from here to discover the families who have lived on our road for generations, spilling out into other plots as the generations pass. In 1939, this was a road of farm labourers, tractor repairers and dairy men and women. A couple retired here from the city where their only daughter still lived in an asylum, along with a thousand other women deemed lunatics. On another day these new layers of insight would pull me up and out of myself, but today they don't spark anything in my brain. The hole where the name of our house should be gapes wider and starts to feel like a conspiracy that sucks me in.

I feel a surge of rage at the 1939 Register, at myself for spending time on this pointless activity, at needing it, at needing the stupid therapy, at thinking we could live here, be parents, stay married and grow our own fucking broccoli. I pick up the books, notepads and magazines – all the research materials I have been collecting since March – and throw them at the wall, the hardbacks leaving dents in the plaster. I tear flimsy print-outs into as many little crumpled pieces as I can and then I look at the computer. Expensive, essential, bought with money I earned from overworking in 2017 when it all went very wrong. I am so tempted to pick up its screen and throw it against the wall, to hear it shatter and to see the chunk of plaster its corner takes out of this ugly room fall to the floor. If I could pick myself up and lob my body at the wallpaper and revel in how broken it is as it slides to the floor I would do it. What an ugly mess. What a stupid, ugly, fat baby of a mess I am. I slap my own face as hard as I can and then do it again and again. The slaps become punches that I hope will leave bruises. Why am I such a mess? For nothing, for no reason. Because I am awful. Because I am shit, pathetic, spoiled. Nothing. I am nothing. A mask hiding a great big void. A human cover-up trying to be perfect – what a joke.

I have the smoothness of the computer screen in my grip, and I could do it. I want to do it. To give up – make an excuse and not follow through with my plan to leap and dive back in to the good and the bad. To smash it into the window and hope the momentum carries me to the moment of collision too. I look up, considering it seriously, and my reflection looks back at me. The sad sight of her makes me pause. Red-faced, wild-haired and about to destroy her way of making a living and the thing she uses to try to understand her world – just because something wasn't in the place it should be. I look past this self, mirrored in glass, smeared across the autumn plot beyond, and I try saying the opposite of all the awful things I have been telling myself. That I matter, that this matters and that there are nice things about me as well as gaps.

After a few minutes this helps a little, enough that I can begin to see past myself to the view outside and the tomato plants, succumbing to blight, done for the year. Next year though . . . next year. I take that sight, the thought it prompts, and rather than let it jump to all the places it could, I corral it into a specific action – one that will give me something rather than take it away. I put the tomatoes between me and the impulse to dash myself to pieces and type 'open source tomato seeds' into the search engine and let the fruits fill the screen and the dark spaces.

Fifteen minutes later, I am lost in hunting for next year's crop – keeping away from thoughts that we might have jacked all this in by then, fixing myself on the screen and not the shredded paper on the floor. After bookmarking some possible varieties, I come across a cultivar called 'Paul Robeson', 'a big dusky red beefsteak Russian heirloom tomato with cult status in the US'. The description of its complex juiciness and 'luscious smoky sweet and tangy flavour' makes me salivate – even though raw tomatoes actually cause me to gag. It looks beautiful too, this fruit with a skin of dark and burnished red–green, and it is what I need today – what I will need next year. I try to add a packet to my basket before moving on to search for garlic sets and discover that they have sold out. Keep focused, I tell myself as the bitter little voice pipes up. I quickly type 'Paul Robeson tomato' into the search bar to find another supplier. The screen fills with pictures – some of tomatoes, yes, but others of an African-American man from the 1940s or 50s maybe. I look at the links below and the first is entitled 'Why Soviet Russia Named a Tomato After an American Celebrity'. I bite and it is, indeed, juicy.

In 1898, when our plot was just the eastmost strip of the pasture called Church Hams, Paul Robeson was born in New Jersey. Paul's father was of the Igbo people but born into the brutality of slavery. The decades between his birth and Paul's had changed the world, though not, it would turn out, nearly enough. As our house was being built, Paul graduated from law school but his legal career was cut short in the face of the racism he endured in everything, every day. The young man turned to acting and singing instead, his wife Essie bringing in the dollars that got them through the harder times, until his performances on stage, screen and concert-hall stage earned him celebrity status. It was then that his commitment to advancing civil rights rose up further and somehow this is where tomatoes come in.

The Paul Robeson tomato isn't complex only in flavour. It houses the story of how Paul craved the freedom and equality missing from the racist society that wanted to take his entertainment but hold him down. Inside its flesh are the visits he made to Russia, the hero's welcome he received there and his love for the motherland, thanks to its promise to dismantle the systems under which he had suffered. There is a simplicity to the story of this fruit's name. Someone in Siberia, and no one appears to know who, bred the most delicious tomatoes that grew perfectly on their land. They acted as a buffer to the poverty and hunger that was supposed to be a thing of the past, but wasn't. They loved this new variety and as they loved Paul Robeson too, they named it after him. There are a thousand complexities in this tale too – contradictions, hypocrisies, brutalities and disappointments. The two-tone of the fruit's skin: hope and ideals mixed with injustices, genocide and corruption. Both stories are true, both are false and each extends far beyond the name of a single plant.

By 1989, as our house's previous owners carved the date into the mantelpiece beam, Robeson had been twenty years interred in Ferncliff Cemetery, New York. His tomato had been growing on quietly in Siberian gardens; producing fruits year on year, keeping people alive – with more than food. And so, two years before the boundaries of the USSR dissolved, an American seed-saver travelled to Siberia in search of at least one of its famous tomatoes. He came back with sixty varieties smuggled about his person and a lasting impression of a society where keeping and sharing seed remained an awful and wonderful necessity.

A year after this trip it snowed on our plot in a big way. Flakes fell in Washington too as George H. W. Bush vetoed a Civil Rights Bill and, in the last months of the USSR, pere-stroika – the Communist Party's radical reforms – allowed Marina Danilenko to start a business with her mother. The collective farm was dead, but she was alive and quick enough to become the owner of the first private seed company in Moscow. She gathered them from farms and backyard growers, selling half a million packets to eager Russian gardeners. The following year, as the Iron Curtain was finally lifted fully, Marina trav-elled to America with 170 varieties of tomato seed, including a packet with 'Pol Robeson' handwritten upon it by her pensioner employees. A gift from a regime that had spoken of equality yet delivered the opposite to a country that talked of freedom but had, for Paul Robeson, been built on captivity.

This is the tomato for me, because I want to grow things that have come from a place of need as well as desire. To plant seeds that have been shared, exchanged and passed along, not just commodified. I want to grow a tomato that is complex – I don't really care about the flavour – I want the history, the nuance of it caught up in the fruit, a reminder and a prompt. The learning is what I'm here for and the beauty, the stories and the names: 'Czerno Krimski', 'Siberian Giant', 'Azoychka'. Hours pass. Surrounded by scattered books and plaster chips that have become invisible to me, I continue to follow the tomato vine back from Russia and try to pronounce words of Nahuatl, the Aztec language: 'tomohuac' (swelling, roundness, fatness) and 'atl' (water) that became 'tomatl'. I see the Spanish discovering a 'new world' – an old world, of course – and turn the fruit they claimed to have discovered into 'tomate'.

At the end of the afternoon, I leave the room. I put my hand on the wall of the corridor feeling dazed and lost in ruby spheres, missing links and black holes – unsure of which is which. There is one thing I know definitively though – that if we stay here I will grow tomatoes. They will keep me pointed away from myself and act as an antidote to the punnets that sit on the supermarket shelf thanks, sometimes, to the horrors of modern slavery.

It's 10 p.m. and I'm back in the damned town car park for the second time in twelve hours. Despite the day I have had, Jared and I have been out for the evening, not wanting to cancel this weeks-old plan or the babysitter we'd booked. I am realising now though, as the glue on the duct tape I'd used to hold myself together earlier loses its stick, that it wasn't the best idea. I'm still upset and as Jared drives us towards home I burst my banks in the supermarket petrol station: a river flowing into the footwell and on to Jared as he gets back in after filling up the car. If this is who I am – I say between sobs – a diagnosable perfectionist, then I can't trust myself any more. I can't trust that I will choose the right path, or even the right tomato. I can't make the decisions I need to make because the model I've been using is faulty and I don't have a new one. Yet we can't wait. We can't live in a perpetual state of suspense until I get my shit together and I won't be able to do that unless I can see a way forwards. It's an impossible puzzle that circles my head and comes out disjointed when I try to explain. I can see I am losing Jared. He wants to comfort me, but it's not something he finds easy. This kind of crying makes him shrink back and then I cry some more.

I tell him through in-breaths that don't reach my lungs that this is all my fault. This house, this smallholding that we cannot afford and do not have time for. I thought it was a reasonable goal: good for us and good for the world. I thought tomatoes were political then, I know they are now, but I can't be trusted any more. I am afraid to repeat the same mistakes again. I can see that the first three of these things were enough for Jared but it's all coming out and I can't turn it off. It is my fault he is depressed. It is my fault our marriage is struggling and my own fault that I feel like this. I did this to us and yet I am still trying to lead the charge and get us out of the mess. I should not be in front any more. I should be missing in action but no one else is stepping up.

I subside eventually and we drive back in silence, a favourite CD filling the gap where I want his voice to be. I hide in the garage until the babysitter has left, so she doesn't see my face, which is more crying than skin, and listen to the tawny owls calling to each other. Her 'to whit' is followed by his 'woo-ooooo' every time. I can still hear their voices as I lie in bed later, each shout seemingly identical to the last, but no less urgent for it. I wonder how close they are perched to each other as they hoot their songs through the dark sky. Jared is right next to me, sleeping, and though we are almost as close as two beings can be, the owls sound closer tonight. I think that I have been calling out to him, but as there's no response I must be doing it wrong.

In the first hours of the new day I wake feeling sick. The rest of the very early morning passes in an up-and-down from bed to bathroom where I sit with my back against the cold porcelain of the basin. It is now just after seven but I am exhausted and so Jared asks Sofya to do the animal jobs. She runs back in to tell me of the first frost, her cheeks pink with cold, her thick winter coat still washing machine-clean. Then I sleep: on and on, all day and through into the next night. I can't move as I sink further down into a thick, dark place. It's warm in here and even when I try to open my eyes they close again by themselves. There's no need for food or drink. I'm a stone covered in moss radiating the heat from the day's sun into the comforting darkness.

On the few occasions that my consciousness bubbles up, I remember that this is exactly how I used to feel as a teenager.

My mother called them my 'blonde days'. I would seem ill and, despite not having any concrete symptoms, would spend a day or two in deep sleep, barely moving. I needed rest so much that my body used to turn itself off periodically, just as it is doing now. The bed is tugging me down again and, before I let it, I try to hang onto a thought, to put it away somewhere safe to come back to later and examine properly. Even with the three brain cells that are still working, I know that I have some important questions to answer. Why was I so exhausted as a child? My 'blonde days' – why did I need them, why do I need them still? I try to carve these notes into the central pillar of my mind as the blackness invades again and my teenage self takes the reins – the adult relinquishing control for a few hours more.

My hair is short. Just seventeen and I've made it to the front desk of the youth orchestra. I'm sure the conductor knows that I mime all the hard passages – and most of the easy ones too – but the players behind can follow my accurate bowing technique and so it doesn't matter that I am only playing the air half a millimetre above the strings. My contribution to the beautiful layers of sound filling the concert hall is a smile and a pantomime of arms. And I enjoy it. As long as everyone else keeps playing their parts, and I pay the closest of attention to absolutely everything, no one will be any the wiser.

The morning after the concert I gallop a horse over the Birmingham country park where I spend most daylight hours when not at school – the nights now taken up with boyfriends, parties and A-level revision. The late-spring air rushes at my face and I eat at least three flies per quarter mile. I don't care: this is a place where all my jangling pieces knit together.

We slow to a walk and take the path that traces the Runnel that – if I knew how to ask the question – could tell me a story of the winter water meadows that were here before picnic tables and local-authority signs. I feel light and full of potential. The drop of my mother's shoulders as she disappeared happily into the garden taught me to take myself outside to feel whole, as did the contented clarity of my father's expression when he returned from the mountains – cuts on his hands from the rocks, blisters on his feet from the boots. This is why, after the hectic week that lies ahead, I will be back here next week-end, the frequency of my body's vibrations slowing as the leaf dapple hits. I'll feel brave enough to see if I can jump the horse right over the Runnel and land on the other side. I'll steel myself, collect her to a canter, prepare to fly over the water and then find myself head-down and soggy in the ditch. From its base I'll hear my friends' hysterical laughter and I'll smile with them because this is a place where it doesn't matter if I screw up. I'll pull my head out of the brook and push up to standing and then they'll stop laughing: silenced by a criss-cross of stone slashes across my right cheek and eye socket. But I'll tell them I am fine and I will really mean it. This is the kind of injury that feels like a badge of honour, each scratch a sign of membership of a club I want and need to be in.

When I was this child I knew this stuff without trying. I saw the simple truth that the land was a place I needed to come back to again and again to balance the effort that the rest of my life seemed to require. When the effort became too much, I was attuned enough to my body to follow its instructions to stop, sleep and switch off entirely. And I understood that there were lessons here and that they would come at me sometimes with the soft fragrance of the hedgerow's rose petals and, just as often, with the insistent stab of its thorns.

October is nearly done and though the tender border plants have survived the first frost I know they don't have long left. I have been tracking the weather closely; the daily temperature forecasts a gardener's friend in autumn and spring. My overwintering seedlings are housed in two plastic rabbit hutches to keep them safe from cats, birds and footballs and the deep trays allow me to bottom-water them as needed. It is very makeshift and today, after heavy rain, I spot a big flaw in my system. Some plants are thriving in the damp but the geraniums hate it, as do the echinacea, and they have expressed their hatred via almost instantaneous death. These trays are going to need to be laboriously emptied of plants whenever it rains, the water tipped out and then everything put back in if I am to stop more of them heading for a watery grave.

There is no way I will do this regularly for the next six months, so I pick the plant housings up and move them inside the ramshackle greenhouse, thinking that 50 per cent glazed will keep more rain out than no glass. We have done nothing about its half-built status since the arguments of spring and it's another tender place that we skirt around. I very much want everything to be fixed, I very much want a greenhouse with all its glass in or a polytunnel, even a cold frame would be a start. Yet there's part of me that revels in the inventiveness that a lack of kit and my inexperience can provoke. Solving a problem by letting my thoughts spread out instantly around it – propping one end of the hutch up on bricks to create a deep end for periodic watering and a drier end in which the seedlings can live – gives my brain a kind of contented satisfaction. A professional set-up would be made of parts that fitted together perfectly for their specified purpose. Mine is a bodge – like the emergency goose-house door that I fixed with a bar that slots into push-in clamps that I found when looking for hinges we didn't have.

As I put the last hollyhocks back in the tray I hear a carrion crow and see two flying over the field towards the village, their shadows brief against the grass. There's a silhouette too – just for a few seconds. A woman maybe, and though she's turned away, there is something familiar in her movement. This figure feels different to the others I have met here; those who were missing and needed me to fill in their gaps. And as the crows' calls fade and the indistinct outline does too, I have a hunch that it isn't up to me to find this shadow and bring her to life. This time I think she is the one hunting for a missing person and that the person she is looking for might turn out to be me.

# I AM REVEALED

I'm working on the plot despite feeling very November-ish. The days have closed in like little vices and the darkness has expanded to fill hours that in summer were daytime. Everything is a rush again and a fight against the night. The first hard frost came in last week too and, within a few days of suspended silver mornings, the dahlias and nasturtiums – all of the annuals that had been beacons of colour against a general slide to grey and brown – turned to mush. I must pull them up and dig out the dahlia tubers, but not today.

I know that this is the time of year I should be finishing off the outdoor jobs that were started in September because any work I do after the intensity of spring and summer and before the ground turns hard will be a gift to myself next year. But I had neither the time nor the energy to begin most of these jobs then and now they hang over me on a twirling mobile of guilt.

There is one project that I am excited about and have not fallen behind with either – I haven't started it yet. Today, I have decided, is the day I will begin work on the meadow, planting the wild flower bulbs that have been waiting in the shed until the ground was soft enough. These small shrivelled things in paper bags will, I hope, give me native springtime flowers like snakeshead fritillary, wood anemone, bluebell and wild daffodil.

I'm planting non-native flowers too: large crocus, nodding star of Bethlehem and camassia, though only because I ordered them in an excited flurry and forgot to check where each originated. Choosing native species was an important part of this project and I am frustrated with myself for rushing and stuffing it up. But they are here now and I will plant them anyway.

I'm using my new hori hori, a Japanese tool with a curved blade – like a long, narrow trowel – serrated on one side and, in this version, with a handy scale of depth marked on. It is already my favourite thing: light and strong, ergonomic and sleek and there's almost no job I can't do with it. Around the grass I go, scattering a mix of bulbs at random and driving the hori hori in where each falls. With a wiggle left and right I make an opening of the correct depth, put the bulb in roots-down and press the clay back together to seal it.

There are hundreds of holes to be made and so this will take me a few sessions. Once I'm done, I'll start the next phase: using a rake or, better still if I can stretch to it, a rented scarifying machine to expose the soil under the thick grass. Then I'll scatter the meadow-making seeds: yellow rattle which will parasite on the vigorous grasses until, given time, they recede – leaving space for a diversity of plants to self-seed. The vision of this future meadow keeps me going even though my hand is aching from gripping the hori hori tightly, my back and knees are sore from crawling and my clothes are wet and muddy. November is not my favourite month, this is not my favourite time in my life, I didn't even mean to buy some of these bulbs. But I am doing it anyway, as if instructed – and I am enjoying it.

The last bulbs for today are camassia. I plan to bury these a few inches under the soil flanking the path, in the hope of making a walkway of their flowers – a line of bridesmaids in indigo dresses throwing their petals at us as we walk to the front door. Despite not originating here, but in parts of North America, camassia are loved by a wide variety of Kentish insects who don't care where the pollen originated. I spent last night digging around in this plant's history, enjoying the richness of its journey to me, and so I now know a few of the many stories hidden inside these little white globes covered so shabbily in brown paper. I can't stop hankering for knowledge of the world and for a chance to break out of the difficulties of now. I want to feel as though my hand and arm are elsewhere and so I close my eyes and wait for the next current to direct my course.

The women carry wooden sticks and baskets as they walk across the woodland to the open meadow beyond. They have set up camp near this food-gathering place. Generations past are busy at their elbows and heels, tutting and guiding them as they work. They will return again when the leaves start to fall so as to divide, harvest and replant the rhizomes of t̓axwt̓ak̓wa̱s – the springbank clover – which is now abuzz with anthophora bees. Through the green of late spring I see the purple blooms of the muṯa̱xsdi; the camassia. The turf has already been lifted and one of the younger women is easing the bulbs up from a stubborn clump of roots in which they are caught. Some of this harvest will be roasted in a pit, turning sweet as it cooks for more than a day. Another portion will be dried for trading or stored for marriage negotiations to come. And, as the women of Tsawataineuk First Nation people have always done, the bulbs – no larger than a kwaskwas', a bluejay's, egg – are divided and half are put back in the earth. So much of their community's safe passage, systems and happiness are bound up with these treasures, so they work to regenerate the meadow every year, securing its future and the future of the crop.

I feel my hand still on the hori hori as I search for the mutaxsdi. I am trying to get closer, to merge with the women working the meadow, but something is in the way. I make another attempt to connect, but it is no use – my fingers won't meet the stick or the flaky surface of their bulbs. This is not quite who I am or where I am meant to travel. Eventually I stop struggling and admit something to myself and sink into where I really am in this scene.

In one hand I have a wooden stake and in the other a mallet. This is not a harvest day for me and I am not following an unbroken line across the land – I am putting up a fence. It is hot today and I am thirsty, so I call to my son to bring the water to me, drinking it in big gulps and splashing a little on my already wet face. This is hard work and lonely too, but no one said it would be easy. The ship, the hunger, the walking, finding this land and making a new life on it. Settling here and claiming it – because no one else has – because we need it and because we can. It will be worth it.

I bash another stake in, and quickly this time because those women are at the edge of the river again looking as if they want to start digging everything up. I wave my arms and shout – who do they think they are – and, as they stand in the distance looking back at me, I call for the dogs and get my gun. The sight of their strange clothes and steady stares gives me a nasty feeling in my stomach and so I fire a shot into the air as the sound of barking travels towards them and they begin to move away. I will run cattle on here presently and their hooves and jaws will soon see to the flowers these trespassers seem to want so much. It is a shame, I let myself think it for a moment, to lose this purple carpet that spreads as far as I can see. I go out beyond the fence and spend a little time easing a few of the bulbs out before it's too late and plant them next to the house to remind me of the bluebells and dog violet I have left behind.

I snap back to this not-yet-meadow, my hori hori and the bulbs that I didn't mean to buy. I cannot see through the eyes of the women of the Tsawataineuk First Nation people and I definitely can't slide into them as I have with others. Still, surrounded by the sounds of the river, the breeze blowing through the meadow and the smell of the muṯaxsdi roasting in the fire, my senses were pulled towards a vision of who I want to be, who I wish I could be – a guardian, a custodian of the land. I want to understand the reality of other peoples and not simplistic, reductive versions full of caricatures either, but I can't do it by pretending to be something I'm not or denying the past. I am the bad guy in this story: the white woman putting up the fence, breaking something powerful I didn't even try to understand. My part in the camassia's tale is brutality – a placing of myself and my family between the harvest and those who ensured it was there for me to take. There will be many stories that are not mine to live or to tell, and learning to recognise that might be the first step.

Ben Cherry is trying his best today, as ever. He wants to move us on from thinking about the overarching problem he believes

I have – clinical perfectionism – to looking at the specific ways it has infected my life so that we can begin to tackle them. He has drawn something called 'a worry tree' and we are laughing like old friends about how CBT loves a diagram with arrows. 'Arrows?' intone the CBT gods, 'That'll fix 'em! Draw a few more!' I settle myself down as he explains his diagram to me – a process for dealing with worrying thoughts. At the tree's tip is the worry and, after noticing it, I am supposed to follow the arrows through the branches and down the trunk, de-escalating my panic by finding out what it is I am anxious about, whether it is real or imagined, if I can do anything about it (and if so when) and then make a plan of action with one thought flowing into the next. There are a few useful things hanging from this sketch's branches but mostly it is giving me the urge to laugh. I try really, really hard not to – these sessions are gold dust and this man is one of the golden people in my life right now; trying to help, giving generously of his effort and time. A giggle pops out despite this effort and then I shake my head as I smile. The younger man looks at me quizzically and I look right back at him. 'Ben,' I explain, 'I don't need a worry tree – I am a worry tree.' But he smiles then too, a laugh before asking me if I can say a bit more.

This process you've drawn – I do it naturally, I explain, and about everything, all the time, almost constantly. I will admit that I am not very good at the letting-go-and-moving-on bit, preferring to find as many routes down the arrows and ensure that every scenario has been covered. However, I do know how to process worry, how to make a plan and tackle it and then move on to the next scenario. Perhaps, I say, I need to work on the opposite: hanging out somewhere in the forest's upper storey, ignoring the arrows and noticing the low cloud stroking the treetops. This leads us to where today's session ends: how to deal with Rebecca the human worry tree.

Ben asks me to set out all the mechanisms I have for planning and organising my life and invites me to consider whether they are helpful or if they are tying me into anxious knots. I explain that without these strategies I fear there would be chaos. I have to do all this to keep the tangle at bay. 'Would it really be chaos, though, without all this?' he challenges me. Maybe not. Perhaps my standards are impossibly high and the disarray I am always trying to hold back is just ordinary life.

He can see I'm not entirely convinced, though at least part of me wants to be, so as this session's homework he challenges me to let go of the reins a little, to abandon some of what he thinks might be over-planning and try to rely on myself.

The weather has forced me inside in front of my computer again, where I'm feeling glad that I didn't shatter it in last month's rage, since replaced by a restless, expectant frustration. Settling to anything seems beyond me and, though I often walk to my oak tree and put my hands against its trunk, all I feel is something hard and cold. The shadow of the woman in the field hasn't returned and the richness of voices I have become accustomed to are absent. As the year begins to fold in on itself, and I find myself alone, there is a powerful force turning me towards the old questions: why am I here and what is it that I am for?

This morning, I am continuing my on/off search for a way of describing what I think I am doing here and for others who share a similar goal. So far I have found only August-dry things, like the government's sixty-seventh official annual report into smallholding, which almost exclusively covers quite large holdings. The avenues I take through agricultural policy, farming organisations and horticulture are not the right fit either, and though the world of allotments and city growing spaces attracts me, I can't pretend that they are close to what we're doing here.

Yet finally there is something – the Landworkers' Alliance – a group who describe themselves as a union of farmers, growers, foresters and land-based workers. They are striving for a future in which 'producers can work with dignity to earn a decent living' and where 'everyone can access local, healthy and affordable food, fuel and fibre – a food and land-use system based on agroecology and food sovereignty that furthers social and environmental justice'. Maybe, just maybe, I am getting close.

Food sovereignty is a term I know a little about. With its assertion of independence, the word's hope is that communities will control their food supply and that what they need can be produced sustainably and in a way that keeps them and their cultures safe. Agroecology is a new one for me but its six syllables sound like the beginning of a song I want to sing. I discover that, just as the etymology suggests, agroecology is a way of farming in a deep and respectful dialogue with nature. A few minutes later the words of the alliance's report on small farms fill the screen. I am thrilled to discover that here at least they really do understand the meaning of the word 'small'. One featured case study is just over an acre, yet manages to provide part-time jobs to four people. There's a feeling of something fitting well at last.

The wind lobs another bucket of rain at the window as I carry on reading about how good harvests are vital, but so is the health of the soil and ecosystem. That these should not be seen as add-ons but as crops to be farmed with equal intent and value. Wealth is not purely financial here but includes what the natural world could gain through the act of farming. And, according to these calculations, the people who work the land matter too.

I am so excited by these definitions of success that I have to read sentences repeatedly to fix them down. Then I get to the lines that flood me with relief as well as delight. The Alliance includes non-commercial holdings in their definition of small farms – places exactly like our plot – and advocates for a future in which there is proper support, financial and otherwise, for people who want to do what I am attempting. Time, money, training and equipment would make a huge difference here but I didn't think we qualified or even deserved it. Yet I've always known that there was value in what we are attempting. I shouldn't need a report with a logo to tell me my plans aren't ridiculous. But with these new definitions of what might be wrong with me – too anxious, a pathological perfectionist, and all that has happened over the past few years – I find that I needed the reassurance of this report very much.

Discovering this portal to a world where I feel I belong is thrilling, yet the ideas and problems it contains are far from recent inventions – they are ancient and I know it. And that is how I find my way back to purple flowers and springbank clover. Still not one of the Tsawataineuk myself, but a humble, silent guest, thankful that they are letting me walk in their long shadows and listen as they talk in the warm evening air. There is a lot to learn here and, as I start, I try to remember that these were individuals: very good, very bad and everything in between; humans – fallible, remarkable, ordinary. Their names, their desires and betrayals were as distinct as mine, despite the gap between us in time. I don't know the name of the young Tsawataineuk woman I saw struggling with a clump of roots, but I have now heard of Cheryl Bryce, knowledge keeper of the Lekwungen Songhees First Nation people today. It is she who, among others, is reinvigorating the practice of cultivating, harvesting and cooking Kwetlal – camassia – in the meadows and hills of Meegan, Canada, reforging connections between people and land that were severed, for some, generations ago.

The words on my screen have asked a similar thing of me: to tune in, listen, to harness not destroy. To take time to make the relationship between me and this place solid and strong, because it will be worth it. There is so much more to what I can grow here aside from how quickly it can be turned into cash.

The month moves on and though it has been raining for a week, I have finished the meadow planting between downpours. The sky is grey today but not ominous so I head out with my rake to try to clear the grassy thatch and expose the soil beneath before I broadcast the yellow rattle seeds. Yet after a few scratches it is clear that the tiny window between clay too dry and clay too wet has already closed. This is ooze now and I won't achieve anything other than churning up a mud pit. The 500 grams of yellow rattle seed – a really good weekly shop's worth – can't be sown. I should have been more organised and thought this through instead of charging ahead (like I always do) as soon as the idea entered my brain. To console myself I decide to keep the bag of seed in the cool of the shed over winter, to mimic the necessary frosts needed for germination in the freezer's drawer, and to hope for the best with a spring sowing.

To make sure this happens I start to add a series of reminders to my calendar and my 'to do' app, including a few other jobs in my 'garden' list for good measure. There are other lists on here too: work, admin, birthdays, shopping, children and school, house renovation and more. I am about to flick through them – checking what else was to be done today – and then the thought of Ben Cherry's homework moves into view: shed the shackles and free myself from the lists. Resolved and diligent I delete a couple of the items that might be considered overkill and turn off notifications for my calendar and app for good measure before putting my phone away out of sight where its controlling presence can't put me off my new free-flowing approach to life.

This morning started inauspiciously. The busy month seemed to have caught up with me and I woke up feeling on the back foot, reaching for the lists I'm not supposed to be making and having to stare out at the wind blowing the last leaves down for a good ten minutes before I could calm myself enough to get a grip on the day. But a few hours later I have turned things around and am deep in setting up a new page on my website for a brilliant idea I thought of an hour and a half ago. My brain is zipping and I'm excited – immersed in this new project that is taking me out of the early-winter dark. My phone rings but, because the number is not one I recognise, I don't answer and a minute later I hear the buzz of a text. 'Are you still ok for the meeting? Can't get hold of you?'

I don't remember anything about a meeting, but my stomach goes over the humpback bridge of horror anyway. There's no name on this message so I open my diary – nothing – and search my emails to find a clue. It's only in my recently switched-off app that I finally see the quelled reminder to do something important, with a woman who is paying me, forty-five minutes ago. A sick feeling rises up around a new hollowness – a full body clench and cringe made of shame, panic and a level of terror I can't explain. Part of myself is splitting off and deciding that it might not bother coming back. I try to keep away from how disproportionately awful this feels with tight focus on the practicalities: texting her back with something vague that I tell myself is not quite a lie (but is) about a family situation and phone reception, giving huge apologies and offering to reschedule. I can't admit I forgot and that I am so ashamed of this forgetting that there is no way I will pick up the phone and call.

Thankfully the woman is understanding and so I put a new time in my diary for next week and try to let it go. I have been the worry tree, the worry is dealt with and now I should move on. The brilliant idea needs more work and so I turn back to it but find I can't concentrate, its shine having dimmed in the past few minutes – less stroke of genius and more a way of putting off something boring I didn't want to do. My next move is to turn to my default distraction of flicking through the news. A woman has died because a month's worth of rain has fallen in twenty-four hours and more flood warnings tumble their yellow words of alarm across the screen. The general election campaign is steamrollering on against a background of Brexit, huge climate-change protests and fracking halted by the government with immediate effect despite all the years spent claiming it was absolutely fine.

Quickly it is clear that news is not going to help me at all, so I turn back to the Landworkers' Alliance. Following a link from its site, I discover La Via Campesina which is, in essence, the international union of peasants. Within minutes I've learned what my chromosomes could have told me, that despite often being the last to benefit, women produce 70 per cent of the world's food – when they aren't busy candling eggs that is. Reading about how often the peasants of today are bullied and criminalised, part of my brain is instantly caught up in a riot that took place in my own village in the nineteenth century, when a wealthy landowner tried to take back the plots that had been given to the poor to feed their families. A group of families described as a 'mob' was taken to court and fined for protecting the land they had been gifted against starvation. Though the protesters won in the end, many of their children sailed away from them forever towards the hope of new lives in Australia, America and Canada that very summer. I see them landing on the shore, making the long walk to an open meadow of purple flowers and watch as they start to build a fence.

Through the pages of La Via Campesina I move from the peasants of my village to Ethiopia, Andalusia, South India, Colombia and Tanzania – individual stories, common struggles. It is then that I find my way to the United Nations Declaration on the Rights of Peasants. The dull specificity of the words has me reading with the kind of hunger usually reserved for the last pages of a novel and by the time I get to the end of page four, I am crying as I read the definition of the word 'peasant':

For the purposes of the present Declaration, a peasant is any person who engages or who seeks to engage, alone, or in association with others or as a community, in small-scale agricultural production for subsistence and/or for the market, and who relies significantly, though not necessarily exclusively, on family or household labour and other non-monetised ways of organising labour, and who has a special dependency on and attachment to the land.

I haven't known who or what I am for a long time, forever perhaps. The past three years have been a lurch between an obsessive conviction that I am on a path to something I need to do and an equally compelling belief that I am fraud. I have been absolutely fine and also mentally ill. I have been a normal person, someone who needs to see a psychiatrist as soon as possible, someone who doesn't need help after all. Living an idyllic life, having a terrible time. I still don't know which parts of which of these things add up to the truth, but I am now absolutely clear that according to the United Nations I am a peasant. I might be one of the most privileged peasants there is but peasant I am and this is the place I belong.

The end of the month brings a day of no school for the children and a bird's nest down onto the gravel of the drive. The four of us have spent a happy hour examining it and trying to work out what bird it belonged to – a mistle or song thrush we eventually decide. The nest is a masterclass in haphazard construction – the mud I have been complaining about turned into an art form. We marvel at it: moss, paper, grass and even the blue of a cheap plastic bag and listen to a recording of the mistle thrush's song.

I want the children to grow up recognising the sounds they hear as individual voices, not just a monolith of tweets. We protect the things we have relationships with and they are going to need to be protectors of this world. When we have committed the notes to memory, Sofya and Arthur run off and I set the nest aside on the ground. I feel content, like I might be a good mother, and I am letting that rare thought flood me when I pick up my phone – silenced so I could concentrate on our outdoor fun. There are three missed calls and two text messages. The first reads, 'Hoping you are still ok for our rescheduled meeting?' I can't bring myself to read the second.

Fuck! I say it out loud. The meeting I missed last week was rescheduled for – oh my god – an hour ago and I have forgotten again. It was in my mind this morning – the reminders set and I even thought about it at lunchtime, but then we found the bird's nest and it vanished. The sick feeling has returned at double strength and I try to push it down, along with the empty dread spread across me. I tell Jared what's happened and ask him what I should do and before he can answer, Sofya, who has been listening, pipes up with a memory. Do I remember the time I forgot to go to her school play and how she looked for me in the audience but I wasn't there? And how she was in another play three months later, a chance to make amends. Yes, I remember.

I had been thinking about the second play every day that week, crushed by her serious little face asking me not to forget this time. As she went into her classroom I wished her luck, told her about 'breaking a leg' and said I would see her that afternoon. But it was Friday and Fridays were the one day where I didn't do anything apart from look after two year-old Arthur. So once Friday got going, my thoughts were only on the relief of having nothing to do and nowhere to go. The school hall was in an inaccessible compartment of my brain, only unlocked at pick-up time when I saw children coming out in costume, holding hands with proud-looking parents. I slammed on my brakes instinctively – an emergency stop in the road seeming the only appropriate response. For a second time in three months, my six-year-old daughter had said her lines, hopped, skipped and tried to smile while she scanned the seats for her missing parent. We both cried all the way home.

But, I have told myself, this was just an aberration. I have never forgotten a play or assembly or parents' evening before or since – let alone two in a row. I am organised. I don't forget meetings – not one, never two – because I put everything on my list, mark it in my calendar, my reminders, write it in marker pen on a Post-it note if I have to. These lapses do not happen to me because I have systems to stop them and, usually, when a nice man isn't drawing arrows that instruct me to let go, I hang on tight. In the event that my knuckles are too sore from clinging on, I have a back-up system. The bullet points and diary entries run in a constant ticker tape across my thoughts. I don't forget because I hold it all in my head – everything, at all times – and try really hard never to look away.

Blonde days and ticker tape: a result and a cause. I pair them up for the first time this afternoon as I shut the field gate. The sun is setting – a full seven minutes before four on this last day of November – and it is now that the month's real revelation comes to me in an indistinct whisper. There is something important in this forgotten meeting. I have already discovered new aspects of myself this month – that I am a villain in some stories and a peasant in others – but there is something bigger to come.

I collect the bird's nest on my way inside, the sky dimming gently behind thick cloud. It has a hole in the bottom, concealed by a thin layer of moss that makes the base look solid when it is not. However many hours the thrush spent weaving the mud, the roots and ribbons of plastic together, this basket would not have held everything it needed to. One little egg might survive, two even, but the weight of the four or five eggs that these birds are compelled to lay would have split the flimsy platform, made known the hidden hole and turned those futures into a splatter of yolk, white and shell on the earth that finally broke its fall.

# THE MIND SCRAMBLER

The waning half-moon is still shining possessively over the plot at 7.30 a.m. today as I put Jared's coat on over my pyjamas. I choose my path over the lawn carefully, trying not to wear a track in the same place each time. I fetch the goats a couple more segments of hay, patting myself on the back when, for once, the baling-twine scissors are there on the hook that I screwed into the shed ceiling for just this purpose. I have managed this tiny thing at least, at a time of my life that otherwise has not gone exactly to plan.

Since December arrived, every ordinary step I take feels full of possible meaning and each footfall on this wet grass equally laden with the possibility of making a muddy hole. This is the last month of a year that was supposed to see me coming back together, a smile on my face and a new plan for what was ahead. The close of the decade I became a mother – twice – changed careers more times than even I can remember, made two big moves away from friends and homes we loved, had more joy but also more distress than ever before. It has been the biggest, busiest, fullest, emptiest, most beautiful and awful decade of my life so far and I wanted this year to play out as a fitting denouement. Three hundred and sixty-five days of resolution that would lead me out of this adrenalin rush into something calmer, deeper and more stable. It has not turned out that way and I know that I am grasping at these last hours and willing them to deliver a Christmas miracle.

Meanwhile the puddle at the field gate has become a real pond once more. Rain and earth – I thought I knew all about how they mix but the ground this December is the wettest I have seen. Victor has lived here for forty years and even he is shocked by the rising water. Last week our plot's former owners visited to find the road turned to river, the lawns a series of lakes, and the east-facing flower beds nearly a foot deep in flood. In all their thirty-three years here, there hadn't been a winter like it. It's frightening: the pace and suddenness of the changing climate evidenced by mould growing along our skirting board and ducks swimming happily on the carriageway.

I like my winters cold and metallic – hard, blue and fractal – not soggy and mild. There is beauty to be found in this sludge, I am sure, but it's not obvious to me. There is pressure – though only from myself – to draw a line under this time. But I don't know what it is inked in yet and very much hope it doesn't turn out to be mud.

I have stayed out here this morning to harvest some vegetables. Sofya asks for honey parsnips every Christmas and so I sowed these in April and have been working on them ever since. The post-mortem on this year's vegetable garden shows an occasional triumph or two, a patchwork of outright failures, some mild disappointments and a few good starts tempered by later neglect. It didn't help that it was too cold and then too hot, but I can't blame it all on the weather. I put the energy I had into flowers more than food because I needed to look at beautiful things. After expending any motivation that remained

on dragging myself out of June, there was nothing left for pot-
ting on peppers and feeding the squash. The tiny segments of
time in which I allowed myself to live this summer were the
product of real need, but living in the moment does not make
for a diligent smallholder.

The parsnips, however, look as if they have worked. I am
kneeling as I start planning how to get them out of the clay that's
holding onto water as if it remembers the summer drought.
From this lowered perspective I can see the plot's swampy surface, weeds, the skeletons of frostbitten calendula and a world
of discarded pots, labels and canes. My vegetable garden feels
like a town evacuated in disaster and then abandoned.

I excavate enough soil around the first parsnip to allow me
to grip its top. The ground slurps and sucks at the yellow/white
vegetable, but I persist. Once I have enough of a purchase I
start to twist, a little left and right at first and then in a complete
circle, which breaks the little roots that tie it to the earth. With
relative ease I am now able to pull it up: huge, perfect(ish) and
intact – a really excellent parsnip. I fill my bucket with a dozen
or so and I move on to the line of leeks – the only other edible
I have growing. The rigmarole starts again; removing the soil,
turning the leek this way and that until it is free and I can lift
it with very little effort.

As I set the leek aside, a couple of inches thick and two
foot long – a beauty – I think of my first attempt at a leek harvest – nearly three years ago. I am stronger than I was then for
sure – physically at least. It's more than muscle though: I have
learned how to use my strength and become familiar with these
plants; held their seeds, thinned them and watched as they
grew. Knowing a little of what they are doing under the soil's

surface means I know to rotate not pull. I have the right tools (some of them anyway) and despite the blister appearing on my winter-soft hands, I even know how to use them.

I have always loved Christmas and today I stoke my festive feelings by getting presents ready for when my family visit. I am giving my mother plants this year, wanting to gift something I have grown to the person who taught me about finding peace in a garden. I pot the Ammi majus, larkspur and stocks into larger containers and then move on. Next I fill decorative pots with hyacinth and narcissus bulbs that have been 'forced' – pre-chilled to mimic winter – to flower early. When the last are covered with soil, following instructions from the floatiest YouTuber on the internet, I hide the ugly joins between the inner and outer planters with moss I pulled from the lawn, feeling like a bird in spring.

The Christmases I enjoyed growing up were magical and we continued to spend our festive days in the first years of Sofya's and then Arthur's lives at my parents' house. We went to the same fir tree farm as when I was little and Sofya helped my mother put a golden angel on the top, just as I used to do. There was orange and cinnamon, mistletoe, a magical reindeer who arrived on Christmas Eve and the familiar excitement made all the richer for seeing my own children within it.

This year the celebrations will take place at our plot and with a feast cooked by Jared and vegetables grown by me. I hope it will be extra special, but know it is especially fragile too. The first for five years that we will all be together: the four of us inviting my parents and Liam to join us here on our smallholding after a deliberate gap.

Though Sofya's first Christmas brought alpine levels of snow, a new kitten and an avalanche of doting family, it also ushered in a new underscore of fear and watchfulness. Anxious glances at Liam, his odd behaviour, worries and whispers about that were never openly discussed. There was the year of the emergency hospital stay, his suffering, the guilt, surgery and a glass balustrade broken in an attempt to ease a terrible pain. Each December a new hope dashed or a terrible thing covered in holly-berry paper and tinsel.

Finally, the worst year blew in with the north-east wind. A telltale overconfidence of sharp suits and fast rental cars. Elaborate presents, talks of extravagant apartments and expensive plans that set off a thousand alarm bells and made me grip my thighs so hard under the table that my fingers left bruises. We all put in huge efforts to keep the happy, tangerine-scented show on the road, but we didn't quite manage it – the effort of pretending had become too much and there were tears, a big row and bonds partly severed. For the first time reality and the picture in front of me were so mismatched that I couldn't tell if I was moving or standing still.

So, halfway through this decade Jared and I decided to make our own festivities. My parents have joined us for part of the time every year since and have been generous, fun and helpful and we all did a pretty good job of pouring glitter over the place where Liam should have been. Five seasons on, it feels as if it might be okay to try again. My brother is doing much better and so, cautiously, I have opened the door and found someone behind it both to discover and to remember. I keep my eye on the many disappointments of the past but am choosing to find new trust and hope inside nonetheless.

As I sweep up the squiggles of moss, the intensity of all that this celebration needs to contain presses against me and so, when I've finished, I divert myself for a while: emails, the weather forecast and then a click to Instagram. As ever I fall back on the soothing online pattern of following one connection to the next, mimicking the natural rhythm of my brain. A picture with a tag that takes me to a new account. A hashtag that leads to a thought, that sparks a search-engine query and finally a post shared by a friend with a link to a piece in – *The Cut* – an American magazine that I haven't heard of before. I click automatically, as I have been doing for the past half an hour, taking in that the article is authored by ten women writers sharing what it feels like to have attention deficit hyperactivity disorder – ADHD – and not be diagnosed until their twenties or thirties.

My reading is absent-minded at first – another distraction from the feeling that many aspects of my life are trying to resolve themselves over high-stakes turkey and Auld Lang Syne. Because of the work I sometimes do writing about women's health, and a friend who has recently found out that she has this condition, I know a little about how ADHD has been underdiagnosed in women and girls. As I skim the introduction I don't learn anything new. I am about to click away to elsewhere on the site, but just before I do I reach the section where the women tell their stories, the first of which starts like this:

There's this carnival ride called the Mind Scrambler that spins the seats past each other, gradually gaining speed while slinging you out to one corner and then another, over and over again. ADHD kind of feels like that – not in the fun way, but in the way that all you can really do is hold on and hope that it slows down.

I get to the end of this sentence and then go back to its start. I cycle through a few more times, not really breathing while I do. As I read I see the ride spinning faster and faster while someone who looks like me holds on, tries to slow down but finds everything spins more violently instead. Eventually I move beyond the full stop to hear more of the story of this woman who finds the world a smorgasbord of inspiration and importance and says, 'I'm aware of so much that seems interesting and it's really hard to cut anything off.'

Another talks about her energy as 'the best and worst thing in the world'. She is, she writes 'a medium-sized hurricane', adding that it can be amazing 'when that power is directed in a positive way' but damaging and difficult when the energy is destructive and restlessness makes her lash out. She describes being flooded by emotion – entirely consumed by whatever it is she is feeling – and the meltdowns that follow. These are clever, successful, professional women, including a medic who was diagnosed as an adult. She was really bright and tried really hard, got excellent grades and so ADHD wasn't ever considered; meanwhile she suffered and stressed and took herself beyond any normal breaking point, thinking that was just how it would always be for her.

The Mind Scrambler: a fairground ride, a metaphor in an article that has just begun to unscramble something in me. I sit for a minute flicking back to particular sentences and then applying them to my own life. Underneath the story I have been telling myself for decades, perhaps I have been riding this swirl every single day. My bedside table, the desk drawer, the wardrobe (and the other wardrobe) appear like a series of slides behind my eyes all full of tangled crap pushed out of sight. Not because I don't care about mess – I really care – I just can't seem to avoid it, however hard I try.

Constantly feeling as though I might be caught out, discovered or revealed is normal for me and the words of these women are, for the first time, giving me a context for this. The spinning vortex in my head; the systems, the repeated checks and always being early for everything in case – horror – I am late; there are explanations for all these things to be found here. I travel across this new information with my usual speed, already trying to unpick and apply it, and then turning it into a single realisation. Maybe everything feels so hard all of the time because, for me, it is.

A few days later, I am sitting in my office going through photographs of the year that has nearly passed and adding jobs to my garden list. I will have to confess to Ben Cherry that, since I missed those meetings, all my systems have been reinstated. So, when a reminder popped up earlier to tell me to look through the folder in which I've stored the sporadic pictures I took over the growing season, I followed its instructions to the letter. Each photo is marked up with circles and labels – according to some very good advice I found in a book. It is, as I have discovered to the flora's detriment, very easy to forget what I planted and where. Once something has gone over or died back it's almost impossible for me to remember what it looked like or that it needed to be moved. These images are visual prompts: 'move peony – too deep', 'need more white here next year', 'rotate veg beds – potatoes are here 2019'. Following the instructions of each, I am making plans to purchase new seeds, take cuttings and insert calendar prompts.

Like last month's tomato-seed quest, I don't know if I will get to put these notes into action. Jared and I are still circling around whether we stay here and lurch between feeling more together than for a long time and everything going wrong over seemingly nothing. Yet, as I work, I feel calm and held by the knowledge that I am doing this because the me of summer, even in her wobbly state, left this trail to keep next season on track. The first half of the Christmas month is always intense and this is no exception. There are things to remember for school every day, work deadlines looming, more presents to organise, friends to see or put off and a house to prepare for hosting. This methodical job feels like taking refuge from the festive frazzle.

Finally, I reach the end of the latest series of photographs and am jolted from this quiet by the next image in my library: a screenshot – five screenshots in fact. Each captures part of a piece written by a journalist back in June about having ADHD as an adult woman. I have no recollection of ever reading this article, let alone of taking the time to screenshot every paragraph. I did it though – went outside and turned my camera on the west-facing bed so that my future self would remember where the tulips had come up. Read 1,500 words in some kind of fugue state and then turned my camera on that too so that I would find them when I was ready to let their message sink in.

Even before this reminder my thoughts were full of this new, and increasingly believable, revelation. The past few days have been an obsessive frenzy of Googling, bookmarking and note-making. I am familiar with the condition's diagnostic criteria, have read papers about high-functioning people with ADHD, discovered support groups, completed several questionnaires and pored over online forums. That is my way – a laser beam of focus on a shiny new idea that puts everything else into the shadows. Even this tendency itself has become something to be discovered. It is, I now know, not a deficit of attention that's the issue for people with this disorder ('disorder' – I don't like the word but it is so appropriate) but poorly regulated attention: too much of it sometimes, too little at others. The intense concentration that I know well could be hyperfocus, awareness exclusively directed in one place for an abnormally long time at the exclusion of all else. The scattered thoughts, the hitting myself on the side of my head in maths lessons until I could make a number connect to a synapse, it all fits. If I have a brain like the ones I am reading about now, it stops working when it is bored and flicks on like a floodlight at a new scheme.

My brain is welded to this new discovery and I can think of little else. It took huge effort to turn my attention to the garden pictures today and it's a relief to return to the subject my thoughts are really trained on. I read a piece on emotional disregulation and find in it explainers for so much of how I feel. There are trails to follow to other familiar issues: a sensitivity to noise, a lack of dopamine and high instances of anxiety and depression – the mental health cost of the chasm between what is expected of a woman, a mother, a wife, an adult and what someone with this condition can easily do.

I let myself roam this new plain for hours without stopping, not needing food or water but taking nourishment from reading about the wide place up there where my brain always wants to go – the big-picture view that ADHD people naturally seem to have. Other parts of my life take on new meaning too. My need to know every stage of a seed's or chick's development, my love of breaking down words into their component parts and following them back along an etymological journey, this project of peeling back each layer of the plot until I have it in my hand, the particle at the beginning of existence; these things are all the same. A way of trying to track back the enormity of the destination I have arrived at until I have understood the series of logical steps that were skipped in between it and where I started.

This, according to self-help strategies, is what they, we – those of us with 'executive dysfunction' – are told to do. Break it down, make a sequenced plan where one small thing leads to another. That is how you get to the end of a project. Or, in my case, how I get to the end without disintegrating. It feels as if, without knowing why I needed to, I have been trying to help the circuits of my brain's neural networks work in a normal way for years. I haven't ever known about the glitches and deficits in my cerebellum, subcortical structures, or my prefrontal cortex, yet some part of me sensed them. The new waymarkers I have been putting down as my old coping strategies stopped working have gained another layer of importance.

Buzzing with all of this and unable or unwilling to be dragged away, I duck out of as much of spelling practice, bath and story time as I can, emerging only after the children are in bed. Over dinner I fire out all of this new information at Jared for the first time – as calmly as I can, which isn't very calmly at all. It is a lot, of course, for anyone. Not just the change of tack, but how much I already seem to know about it – the pace at which I move. Yet he takes in what he can and agrees that, unlike anything else suggested so far, this really does sound a lot like the me he knows.

We are all sitting at the dining table, underneath a mantel that I have decorated with cut boughs, flowers and berries. There are little vases hidden in this mass to keep the roses from dying and gold candle flame is reflecting in the mirrored surface of the laurel leaves. Though there are plenty of ghosts of Christmas past that could be here today, this 25 December has been joyful. We've made the annual creep to the tree to see if 'he's been', had a chocolate buffet for breakfast and now Liam reads his cracker joke, causing Arthur to laugh the laugh of someone who has never before heard what it was that the policeman said to his tummy.

The meal passes in a warm haze of good smells, happy voices and the occasional sound of a sprout hitting the floor. When we are finished we play charades and I'm in the middle of trying to persuade Arthur that, no, he really isn't allowed to mouth the words to his team, when my alarm goes off. I look up and see the night already off its starting block and running towards me. My animal jobs are such an embedded part of my routine that I don't normally set reminders for them. But today, in case all the festive distraction and good wine scatter me more than usual, I made an exception. With a new awareness of why I might struggle to remember things, I feel a little less conflicted about these systems.

Outside I close the pop holes on each of the henhouses and then walk through the still-mild air to the goat shed where Amber and Belle do not know that it's Christmas, but do mind that their dinner is late. The green scoop digs into their oat mix – and I see Robbo the robin perching on the damson branch that overhangs the roof. I scatter a little feed on the floor for him, hoping that the supposedly goat-perfect balance of ingredients tastes nice to my red-breasted friend. I walk back to the house the longest way possible, enjoying these solitary minutes before returning to the happy fray.

I haven't yet told my parents or Liam anything about my ADHD hunch, but Ben Cherry and I talked about it when we had our final session last week. I described how my attempt to relax my rules had gone. I did not find a middle ground, I told him. The looming pit of chaos was real – full of forgotten meetings, unclean clothes and missed deadlines. 'I think I need the rules,' I explained, letting this be the introduction to the idea that what feels like a drawn-out breakdown (I stammered a bit around the word) could be related to undiagnosed ADHD.

He nodded but was sceptical, having worked, he told me, with children with ADHD – mainly boys. So I shared what I had learned about how women and girls present differently; less obviously hyperactive, less disruptive, less reckless. He nodded again, less sceptical now, especially when I told him that my GP agreed it could well be possible and has referred me to a specialist psychiatrist who I am going to see privately. I do not want another one of those Samaritans letters and so I will pay for help with a credit card this time.

As Ben drew our closing session to an end, I handed him a letter of thanks and a packet of seeds saved from the June flowers that were blooming when I needed help and couldn't get it. He has helped me, though perhaps not exactly in the way he planned, and I wished I could do more for someone working in a system like this and managing to keep hold of their kindness. Then again, he had just shown me the line graph charting my symptoms since the summer – their frequency and severity plummeting from extreme to the safer end of moderate, with nothing but the plot to explain this dramatic improvement. Perhaps the gift of seeds was enough – one of the most meaningful things to give or to receive, as they have always been.

It is dark by the time I am ready to go back inside, pausing first at the window to look in at my family. In the kitchen Jared is pouring drinks, looking relaxed and in charge, somehow energised by making the Christmas dinner, not zapped as I would have been. Liam is talking the children through his present to them, a sketchbook full of plans for a tree house he'll build around our hollow tree. I look at their small heads next to his larger one, tilted towards each other on the sofa with an ease that's only present when there's love flowing in both directions. I can't believe he is here and that it is fine. I can't believe that my brother – this brother – hasn't been here for years, or that his difficult last visit was one of the moments that finally took me from coping to very much not. All this looks different through the lens of the Mind Scrambler too. It took me less than half a day to realise that, if ADHD is part of everything falling apart this decade, the real tragedy might not be my lack of diagnosis – but his.

My father has picked up his guitar, strumming distractedly as he waits until it's time for his role as chief storyteller to be reprised. I love many of the things this man loves – music, books, mountains – and I know I have tried to learn from him too: the calm interested humanity in his voice as he reassured his patients, using only skill and a telephone. As I watch, half hidden by the roses and the clouds blocking the moon, my mother comes into the room. She tweaks the mantelpiece arrangement, thinking perhaps of all the years that she decorated the house I grew up in – with a skill and complete over-the-topness that I have inherited. She didn't sit down and teach me how to turn holly boughs into magic, but she didn't show me how to plant a clematis either. Yet I want to do these things, and am what I am because of her and him – because of the code my parents passed to me from those that came before them: ciphers for the good and others for the bad.

The last day of the year and I am cleaning out the hens even though I am fed up with them. They have not been laying for months. Much longer than their normal, annual moult and laying break. We had to buy eggs at Christmas – at the same time as buying sacks of chicken food. I have wormed them, put apple cider vinegar in their water, changed their diet, bought them a new house and burned the old one in case blood-sucking red mites were lingering in there. All of this work – and today's barrows of manure wheeled through boggy ground on a rainy day – and none of the reward.

I almost don't bother to check the eggboxes when I've finished but I can't resist that bit of hope and so I slide one open and am rewarded by not one, but four eggs. Better still, I see that there is a first egg from one of the chicks I hatched in the spring, now all grown up. It is gorgeous – a deep olive green, which means that my breeding experiment (crossing a pale-green-laying hen with a cockerel from a line that lays a dark brown) has worked. I am a chicken genius and the hens can have a reprieve. The pressure of drawing the decade together neatly and declaring that 2019 was something other than a disaster feels lessened as I take them inside and shout to everyone to come and look. This feels like my reward for hard work and faith. The kids say 'wow!' and mean it and then argue about whose hen laid which. Jared teases that I've bred a hen to lay eggs that match the colour of the kitchen walls. I kiss him and look and we see each other again for a moment, liking the sight very much.

The day's jobs are not yet done, so I step back into the wind and drizzle, spotting a flash of orange in the narrow west-facing bed near the door as I step through it. I rummage behind hollyhocks sown in August, now good-sized plants that should bloom next summer, and find a nasturtium vine still in flower. Its ends are dead and there is frost damage in parts, but most of it has survived in the shelter of the house.

I break off one of the flowers and put it in my mouth, soil and all, and it is as delicious as July: sweet and spicy – a treat. I have earned this taste of summer and am going to enjoy it, alone, because I deserve this moment, and no one else here likes nasturtiums. I stand outside and eat every last flower from the vine before moving on to strip the plant of each leaf as if gorging on the final layer of a box of chocolates hidden under the bed. I make a promise to myself to try letting the new year unfold as gently as each nasturtium dissolves. The sensation of fragmentation, of unravelling, is still here. The beginnings of an understanding of why and how does help, but it hasn't made it go away. And with the world such a churning mess, maybe it never will.

Almost everything I read, every conversation at the moment, is peppered with a particular set of words: unravelling, unspooling, untethering, unhinging and unmooring. I recognise both the feeling and the phrasing, but hearing them from others prompts a new question, one that lives in that bit of me that is always trying to break things down. If we are all unravelling, what is it that we are unravelling from? What were we supposed to be tethered and moored to or spooled around in the first place? What was the other side of our hinge screwed into? And when were we last fixed around and against that solid point; how did we come away from it; how can we get back?

I don't let myself stay with this thought for long, not yet, but look instead at the land and find it luminous despite the short, grey day. There are so many self-sown plants and I recognise them because, with a mix of neglect and deliberateness, I let their parents set seed and it has spread in the direction of the prevailing wind. Next summer these will be pale purple poppies, maroon sweet peas and tall umbrellas of celeriac flowers that will froth above the vegetables. My June garden is emerging from December's bog and there's a shimmer of energy around the place too. As the wind ruffles the field maples' bare branches and a carrion crow shouts above, a half-silhouette steps into the side of my vision. I close my eyes and follow her – this shadow I have been waiting for: a woman who seems familiar yet different to the ones before.

I am walking in the dark forest again, following the woman whose shawl is pulled over her head like a hood. She carries a flickering candle in a brass stick, the light bouncing to where I need it, highlighting a tree root here and a rabbit hole there. The bag is heavy on my shoulder and my waistband is tight, stretched out by something tucked within. I struggle to take it out, careful to keep pace with my guide, and as my eyes flick down I see that the heaviness in my hand is a compass. It points north-west as we move through dense places where the owls hoot high and then scrubby open ground where the air is sweetened with damsons and woodsmoke.

At last she stops and, following her lead without question, so do I. It takes a moment to recognise where we are – the smell of garlic under my feet, gravel paths, a lawn above. She moves and the brightness lands on the curve of a stone wall protecting a palm, its branches spidering outwards. There's someone else here too: a little girl of maybe ten or eleven sitting on a log and chatting unselfconsciously to the tree that's now illuminated by candle fire.

The light moves again and I tuck the compass away, place my bag on the floor and tip out the hori hori and the bulbs within. Their paper covers flake to the ground with a promise of purple flowers and I split them with my fingers putting one half of each back into the bag and then making a hole in which to plant the other. Something lost and looked for now shared and then returned.

Crouching to brush the traces of wild garlic from my knees, I glance at the girl still lost in her game and put my hand on the earth of my childhood garden – but only for a moment. It is time to retrace our steps and, as we do, my guide's shadow remains constant, the candle never gutters and she keeps all dangers at bay on the long walk home. I open my eyes again to this cold December day and her outline persists for a moment. A stamp of night-black at the corner of the field before this woman whose face I never see, but feel like I know, is gone.

I fold up the shape of my shepherd like a piece of paper to keep tucked away for dark days ahead, and return to these last hours of this last day. I go into the kitchen to be with the people I love, feeling surer of the decade ahead now that this one has wrapped itself up a little more neatly than I had thought it might. It is as seductive as ever, the hope of order and solidity that has been growing within me since I opened the nestbox. Here is the resolution I have been craving; my paths covered in woodchippings, my hens in lay again, my gardening to-do list completed for the year, my box of seeds alphabetised and the shed meticulously arranged.

Though I have just promised to let the future wash over me without expectation I am already doing the opposite. As I see Jared scoop my miracle end-of-an-era eggs from the counter, I realise I've been turning everything into portentous patterns again and wagging my finger at myself to keep it that way. A mistake as ever, but such a lovely one to make, for those first few minutes anyway, when all appears finished, clear and calm. So though part of me crumples and wants to cry and shout, a newly awakened sliver of my brain tells me it is all for the best when Jared stumbles, juggles and then cries out as each perfect egg hits the floor with a smash.

**missing you,**

The woman in her twenties,
is amazed she's an anomaly. She's

doing fine – if fine means living
breathing, getting married, having

children – without the chunk
that in us normals holds. half

of all our neurons. Her brain,
the jaw-dropped doctors think,

became expert in alternatives,
workarounds, diversions. Who's to say

she's not finer than the rest,
Our   signals

take no chances, walk
the roads most travelled by, while hers,

                    double-joined,
                              are dancing.


                    Tania Hershman,
        from *and what if we were all allowed to disappear*

PART FOUR

# THE NEW YEAR

# ZOOM IN

TEST START

Blue square, red square, red circle, blue circle, red square, red square: CLICK.

Red square? Red square? I don't remember what a square is but I am sure I never want to see one again. I can't think why I have to click this button but I'm certain that I must. I click. I don't click. I click. CLICK. CLICK. CLICK.

I disappear.

I have become earth now. Crumbled to a fine tilth and blown into the cracks of the keyboard: lost between the 'a' and the 's' and the 'd'. I am so fine and so light that I could plant carrots in myself and they would grow straight and true in my insubstantial being. Nothing exists but the squares and circles and they keep coming at me relentlessly, an endlessness of commas, here in the place where I am stuck:

red, blue, red, blue,

The world as I know it has ended. The plot has died off, died back.

A felled forest: white noise and then silence.

And even the swirl of the wind is stopped.

'You're all done.' The psychiatrist's assistant is still speaking but her voice takes a while to reach me because I am hollow and distant. The shapes keep playing behind my eyes, my senses are disconnected and my head empty. The test doesn't feel as if it will ever be finished, yet the circles and squares have gone from the screen and she is already dismantling the equipment. 'How was that?' she repeats. How. Was. That? There is a pause and then the word 'awful' crashes into the room and I am surprised and then embarrassed to realise that I am the one who lobbed it in. I don't want to be a drama queen when the assistant must watch people do this nonchalantly every week. It can't have been awful. But as I look down at the floor, ready to push myself back and give her some room, I notice the right side of my jaw bone is still wet from the tears and my hand is gripping the button-clicker as if it had been welded on permanently. My thighs shake as I stand and I feel faint as I force myself to unfurl my fingers and drop it on the desk. 'I know it's ridiculous,' I say. 'But, yes, it was awful.'

The assistant moves in her jerky way, clearly stressed by the elderly laptop's refusal to connect to a hotspot on her phone and do whatever else she has to so that this isn't a wasted effort. Finally, the computer stops circling and connects. A couple of clicks, a quick scan – her whole head moving left and right along with her eyes – and then the left-hand corner of her mouth hooks upwards in the half-smile of someone who thought as much. 'I don't usually do this. If there's any doubt – you don't want to interpret this wrongly. Give the wrong result.' Her words tumble out disjointedly, but somehow exactly as they should. I was expecting a goodbye but instead she asks if I would like to know the results right now. Something solid, a certainty after all the waiting.

What if it is finally all over? And, worse, what if it isn't?

She doesn't make me wait, this woman who understands the now, now, nowness of life. Within seconds of me clenching my hands into tight fists and confirming that, yes, I would like to know, thank you, she turns and looks at me and says, 'This is very definitive. You are on the ninety-ninth percentile for inattention and hyperactivity.'

I know what this means, but I pause my reaction and check because I am still foggy and this is an important number to get right. 'As in, at the very top of the spectrum? The most inattentive and hyperactive?' I ask. 'Yes,' she nods, turning the laptop screen to me so I can see incomprehensible graphs, scribbles and numbers. 'If there were a hundred people in this room, all ninety-nine others would be more attentive and less hyperactive than you. You would be the least attentive and the most hyperactive one of all.'

It's a finger snap of information that instantly takes me from where I was a quarter-second ago to a completely different place. Over the past year I have got used to finding things out, uncovering them slowly, coming to realisations, but never this fast, this definite and never this significant. The moment of transition, when I move from normal but not coping to something else entirely is a punch. I have my head in my hands and – even though the walls are thin and there are fifteen occupational therapists working at their desks in the room outside, thinking about the cheese-and-pickle sandwiches they will eat for lunch – fat sobs come out of me. I am crying with my elbows, my spleen, every microscopic red blood cell travelling the paths inside me and the hairs on my legs are weeping, even the ones on my big toes that I pretend don't exist are wicking tears.

This drench of feelings hasn't separated out into distinct thoughts but when, after a little while, the assistant asks, 'Is it relief?', it all becomes a bit clearer. A minute ago the ivy vine that had been cinching tighter around me since I was little was cut at the base. When you cut an ivy like this it is not an instant untangling. The ropes of green take a long time to die back and peel off whatever tree they have been smothering. It lingers and turns brown slowly and still stays put until I get my fingers underneath and pull it off, checking carefully for nests as all manner of stuff lands in my hair and on my face. Surely this is what lies ahead and it will be hard, nail-splitting, back-breaking work. This is just the beginning. Even after all the ivy is gone and the trunk feels the air properly for the first time in years, the marks and indentations will remain – some for a lifetime. But the first cut has been made so, 'Yes,' I reply: some of these tears are, in part, relief and the promise of release.

She is still talking, explaining that anyone in the ninety-sixth percentile and above would be likely to have an ADHD diagnosis, but that there's a big difference between the ninety-sixth percentile and ninety-ninth – the percentile where I, apparently, live. I say, the thought forming as the words leave my mouth, that I can't believe I am hyperactive, that I would never have recognised myself as a hyperactive person. She describes all the women she sees whose almost constant movements are tiny, demure, socially acceptable: tugs on skirts, twirls of hair, little twitches that you wouldn't notice unless you were looking for them. The men, she explains, are usually in contrast – moving dramatically and freely – less inhibited and so more likely to be diagnosed as little boys.

I listen to her while roving over this new information and marking up sections to return to later, these notes to self competing for internal space with images of falling, jumping and swinging from one tree branch to the next because I can't not. I can't help it. I never could. My right hand feels strange, still in a loose gripping position with the feeling of the clicker pressing phantom-like against my thumb. But suddenly it is time for me to pay and leave. She makes an appointment for me to see the psychiatrist who arranged the test – this time for a diagnostic interview – and then I step out into the city daytime with every single thing I ever knew loose and clanking inside me like a recycling bin wheeled across cobblestones. Jared drives us back to the smallholding and I focus on hoping that I haven't left any important parts of myself in that place of nothingness where the test never ends.

Storm Dennis has blown in today, the day after the test, and it's hot on the heels of its predecessor Ciara. Denis was my paternal grandfather's name and he was the son of a farmer from Kilkenny who claimed not to have been clever enough to become a vet and so became a doctor instead. Like his namesake, Denis couldn't help being destructive. My granny's most withering looks were saved for her husband on one of the many occasions when the sleeve of his jacket turned one of her treasures into fragments on the floor.

I find it hard to imagine him as a six-foot city GP with four sons, a cigarette always on the go and presiding over an array of vegetables and fruits in an ornate Victorian glasshouse in their Birmingham suburban garden. By the time I came along he had bent himself short and downsized to an eight-foot-by-six-foot garden-centre greenhouse full of tomatoes, each plant named after a granddaughter. The Gramps of my memories wears his third-best gardening trousers, which look as if they need a swift demotion to fourth-best, has a smear on his cheek, thanks to a covert kitchen raid to 'catch up' with a little ham and mustard, and a dewdrop forever threatening to fall from the end of his nose.

I feel the blast of Storm Dennis on my cheek as I open the animal sheds and watch the livestock peer out nervously, unsure if they should stay huddled in their shelters or run out into the harsh freedom of the wind. I coax them out, feeling a little cruel but reminding myself of the field shelters, the thick hedges and the hazel copse they can take cover in. There is much I haven't explored or understood in my own family. Who taught my grand-father to make growing things part of his life? Why haven't any of his sons showed interest in the soil and what they might plant in it? There are things I want to pass on to my children but I suppose, however hard I try, they won't necessarily agree to take them. And there must be many things that I have passed on unknowingly whether I like it or not.

I stack six garden chairs to make a single unit, hoping that their combined weight will stop them from blowing away. A study I read last night found that 41 per cent of mothers and 51 per cent of fathers of children with ADHD received a diagnosis of the condition themselves. Knowing that this acronym of difference between me and many others is likely a genetic hand-me-down makes it hard not to look through the generations and wonder. Gramps? My maternal grandfather, Bill? I discount the women at first as capable, organised, confident and then I remember that those adjectives have been used by others to describe me.

The civil servant is sowing her tomatoes inside, hiding from the storm, but watching it wind the air up into a frenzy that mirrors her own desire to burst out of herself. These future plants are only specks today, yet they remind her of summer. As she tips her watering can she brushes her arm against the tomatoes' future leaves, releasing the warm cliché of their distinctive smell. She inhales calm purpose, meaning and the feeling of belonging to something beyond her own life. But there's claustrophobia in the air too – and longing, confusion – all muddled with memories of looking up at her grandmother's busy fingers at work in the greenhouse. Yellow flowers of vivid memory that always appear with this scent.

Neither the civil servant nor I exist for each other in this early-spring wind. Somewhere in the ground below us though, our lives do intertwine, and they also tangle with everyone who detects a top note of loss in the tang of tomatoes. She finished her report some time ago and it wasn't long after that I read every word, not seeing a trace of her or her balcony of pots within its twenty-nine pages. Neither of us, writer nor reader, is convinced that this document captures what it could, and this realisation is what prompts her to start writing some things of her own. First a letter of resignation that, along with thanks for the opportunity, suggested there were other reports that needed to be written and different questions that she must ask. Then a journal, where she scribbles in messy pencil about why, in defiance of logic or time available, growing her own tomatoes feels like the most important thing. She typed a long email too – to the person she thought would be beside her as she grew things – containing an apology (of sorts) and an explanation of how it feels to be abandoned at the moment of collapse. On a roll then, she worked at a job application, etched her signature on a decree and then pressed send on the enquiry form for a clinic to which she finally felt ready to travel alone.

She puts the final pot on the warmest windowsill and collects the post from the mat. A catalogue, a bank statement and a letter from the council, which she opens first, allowing a tear at what she finds inside: an offering, a chance and a payment finally made against her many acts of faith. She has been given an allotment: just a quarter-plot, a very small piece of land – a small holding. And it is perfect for her, a woman who will come to know its every frost pocket and insect visitor by working each inch with her hands.

The phone rings, she picks it up, gripping the side of the countertop with her compost-stained hands. 'Yes?' She turns this answer into a thin-voiced question and then clenches herself as she listens for the confirmation she's been waiting for. Her neck flexes in a nod and she lets the tears fall properly this time, focusing inwards and finding something else to hold on to. Something where, once again, smallness is the point.

It's Friday and I'm sitting in the occupational therapy building again for my follow-up appointment with the psychiatrist. He is running late and I was early, which is funny given what he might be about to diagnose in me. He wobbles past the window on his bicycle and then appears a few minutes later, still wearing his helmet and fluorescent tabard. He smiles warmly, as he did when I first met him a few weeks back, but as then I still feel awkward as we walk past the people at their desks and so look determinedly at the floor. Building on the hour and forty-five minutes that I talked at him during our first appointment – long stretches interspersed with occasional questions or comments – we review the results of the test and go through the formal diagnostic questionnaire. From the previous ramble and the formal answers of today, the story of the past year, the past three years and the past thirty-seven years comes out in little spirals of disconnect.

I talk about the things that were so much part of our family mythology that I never gave them a moment's thought. My inability to learn to read until late and then being able to do it overnight after a mean comment from another child. The clumsiness and the series of lost/broken/left-behind things of my childhood that provided funny anecdotes that seem less funny now. Then there are the moments I'd forgotten: the clever excuses for homework I couldn't make myself do, the library fines; the emotional collapse when I had to fend for myself at university; the unpaid bills, the clothes bought impulsively. All managed and hidden of course and never anything too bad. Nothing that would raise a warning flag if you couldn't see the speed of my legs increasing just to stay still.

We talk about the ruining of the last year and the intrusive thoughts of scissors and suicide. 'It's embarrassingly performative,' I say out loud, realising it as I say it. He writes that down, saying he can tell I'm a writer, and I wish I could take it back. Yet it is a performance, but not because the feelings aren't real, they are very real, but because dramatising them and turning them into horrible images was the only way I can let the truth begin to break through the person I have constructed – the person that everybody, including me, has thought I am. I was performing for myself too, of course, though I don't tell him that. Acting out the thing I am going to have to do for real when I am ready: destroy the mask, jumping up and down on it until it is powder, and then filling this blank space with who I really am.

'Do you think you would ever hurt yourself? Kill yourself?' he asks, bringing me back from the sound of fragments of my self scattering across the kitchen floor. 'Is it . . .' he struggles a little, 'that you have to show how bad it is? How you are feeling and make someone listen?' 'Yes . . .' I start but trail off as I see the metal triangles that hold the garage roof up and remember how they started to call to me quite persuasively. I remember the interview I read with the psychologist who described the internal chaos and restlessness for women living with undiagnosed ADHD; how the struggle to keep up the facade and meet society's expectations, and the shame when they don't, leads to a greater risk of suicide and self-harm.

He writes it all down and then, after forty-five minutes, my hunch becomes an official diagnosis that will be written up in an eleven-page letter to my GP. Attention deficit hyperactivity disorder: ADHD. Not a borderline case but a raging inability to concentrate without stimulation or supreme effort and a constant need to be on the move. I feel relieved and almost excited by this clarity. A thing with a name and a treatment plan. I can read books about it and retrace my steps through my life as if seeing it all for the first time.

The consultation is drawing to a close when out of nowhere he asks, 'Why do you think your husband is with you?' I pause. I didn't expect this question, though it's one I keep asking myself. I try to brush it off and say, in a quiet, stuttering way, that I don't know. 'He's stuck with me now, I guess.' The flat joke comes fast because not being able to give an answer doesn't feel like something I want to leave in the air to make its way up through ceiling tiles and out into the world.

'I imagine it's because he loves you,' says the psychiatrist gently but with total confidence, as if it was the most obvious thing. He sets that short sentence down for a minute before continuing. 'You are very hard on yourself. Maybe you need to be less so? Perhaps you need to love yourself more, like he does?' I was expecting the lisdexamphetamine prescription he will be asking my GP to fill, but the one for self-love comes as something of a shock.

Storm Dennis has died away and Arthur and I are in the meadow where the bulbs I planted in November have started to flower. Large crocuses are popping up at random, little purple, white and yellow individuals, stranded in the green of the lawn and blown to an angle by the recent gales. I know they will naturalise and spread to make more of a carpet in time, but for now they are comedically sparse, polka dots of colour and I can't stop laughing. No daffodils or camassia yet but they will come and, for now, Arthur spots another crocus emerging and calls to me excitedly. He's careful to look where he steps so as not to squash another bud that might be pushing up through the soil's surface. 'Well done, Mummy!' he says, looking at the flowers and giving my hand a squeeze in one of those rare moments when a child notices their parent as a person who might need a bit of feedback sometimes.

I needed it I realise, that squeeze of the hand. It's been a funny old start to the year. So as Arthur and I head to the south side of the plot to check on the daffodils' progress too, I try to be kind to myself as my doctor has prescribed and not feel guilty about the muddy mess in the veg patch. It's hard not to as I still haven't tidied up the plot from last year's efforts, the greenhouse is still half-glazed and only about a quarter of the overwintered plants have made it this far thanks to neglect. I tell myself it might not matter though because I'm not going to grow much during the spring and summer to come. I might even sow cover crops over the veg plot to keep the weeds at bay, the soil in good nick and simply have a year off. We will hold on to our animals, but there won't be hatches this spring or goat kids the next. This diagnosis might mean I have to face up to being fundamentally unable to cope with the work of the smallholding, to accept that it was probably my disfunction that led me to make the mistake of coming here at all.

I will scale back, adjust, see how the medication helps – if I can ever get my hands on it. We'll see how we get on, Jared and I, and how we both feel – and then we can decide whether to stay or go. As Arthur and I walk and talk I can see that the garden is getting on with spring, despite me preparing to turn away from it. Snowdrops and primroses are dotted here and there again – even in the field where I don't think I've ever spotted them before. The hens are laying in earnest once more and the geese too. I point out the yellow fuzz of catkins on the hazel copse and Arthur, in turn, notices the birds. We get out the binoculars so we can look at them properly. Nuthatches, the usual army of blue tits and, unseen but heard, the thrushes – a mistle thrush. It think it might be the right thing to step away from here, but thinking and knowing are different things. I am not sure what I would do without it, now that I've worked bits of myself into its ground.

As the sky is already getting dark, Arthur helps me do my evening rounds. One of the hens hasn't been coming out of the coop as much and so I take a look at her. She seems okay, but I feel sure there's something that will make itself known soon, old age perhaps. Frisbee is there too, with the look of a hen about to go broody. We watch her for a bit and then I spot Alert, a small cockerel, sitting near a builder's bag, perched in a way that looks unusual somehow and causes a chain of thoughts and conclusions to thread together almost instantly. At my instruction Arthur opens the bag, then laughs and points because Alert's favourite hen is hidden within on a little nest of white eggs, just as I knew she would be. I tell him I have X-ray vision and he is thrilled with me. I am good at this, though it's not the super power I'm pretending. I might miss or forget to pay attention to some key details, but I often spot other things before they happen and am attuned to hidden moments that many miss. I naturally look for and find clues, and bring them together to make a solution before anyone has mentioned that there is a puzzle. It is hard to reconcile these two things and understand how they can both be true.

Before I left the psychiatrist's office for the last time he told me about a study focused on the Ariaal people, a traditionally nomadic tribe of northern Kenyan cattle herders. The researchers tested members of the tribe for the DRD4/7R allele – a version of the gene associated with the brain's dopamine receptors and one linked to ADHD behaviours.

This allele was particularly prevalent in the Ariaal people and when found in those still living a nomadic lifestyle it went hand in hand with markers of success such as better health. Yet, in those who had moved to a more static life the very opposite was the case. I was interested by this so found more studies on other nomadic peoples – and have gobbled them up. One theory suggests that instead of being a sign of a malfunctioning brain, ADHD is part of an evolutionary process that made some people exceptional at looking towards the next day, the next move, and deciding which direction their people should take across the land. They were the gifted leaders who noticed the bigger picture, the patterns within, and drew them together to make the best decision, though not necessarily the most predictable one. But for people who thrived on a changing landscape, the world's movement to a different kind of life could well be full of struggle and suffering.

'It can be a gift,' said the psychiatrist encouragingly, after telling me about the study. 'No one wants to do boring things and people with ADHD can't do them. Look at the choices you've made – your work, where you live. You have made your life around whatever you find interesting and important because you have had to. It can be such a strength if you can adapt and use it.' I believe him, though quite how this fits my life and with being a woman, a mother, a wife, a person who needs to earn a living in twenty-first century Britain, I haven't yet worked out.

I am trying though, which is perhaps why, after Arthur and I run with the greedy goats towards their house and dinner, I tell him a story. And it's one that lives in the gap between the prehuman forest and its first settlers: the story of the nomads. They were the first to come here, to brave the darkness and discover some of the forest's secrets. Some brought their herds of swine with them I say, and make Arthur laugh with a bad impression of the pigs next door. They had sheep and cows too I tell him: shiphurst, cowden, cowlees, they gave these names to the places to which they returned. At first only the men and the pigs came to the dens – the little openings they had made in the forest so they could camp for a while at the end of the long summer. But, as time passed, this transhumance – I pause over the long word to explain it – this travelling to and from the woodland pastures was over and they decided to stay. This, I tell him, is where they made their home. Digging up roots to put roots down. And he nods again as if he can see it.

Arthur runs inside to warm his hands and as I finish my jobs I keep thinking about them and wondering how they fared, the best of those nomads, once they had tied themselves down. Did they struggle like the Ariaal, like me – the leaders falling back now that their brains were forced to work in an unaccustomed way? Or was it a relief? Perhaps they found a way to skate the seasons and the changes in this place as I am trying to, as if each day was a new journey and their challenge, despite remaining static, still to guide their group safely to the next camp. I don't know if some of the nomads who tracked this forest were broken by the constraints of permanence. But I know that their undulating existences shaped our villages, forests and fields. The droves they used to travel over the land still cross it today, pretending to be roads and footpaths. So while the historians and archaeologists put the Wealden nomads in the early Holocene period, I find them in the Anthropocene as well. They are still here now; their move to permanence a very early loop in a thread that takes us all the way to the current environmental crises. I find them in my genes too and want to know which nomads I am descended from and what land I should be walking.

It is the very end of February and I'm nearly home after a five-hour journey from Birmingham. This trip combined work with a chance to tell my parents about the ADHD diagnosis. I spoke to Liam about it all a while back. He has been helpful, supportive, perhaps because his brain has lots of overlaps with mine. I was more cautious with my parents, not wanting to mention this development until I was sure. They have been through a lot and I worried that sharing this odd revelation with its strange acronym containing all kinds of connotations and misconceptions could be another blow. But they took it well and if they were upset, disappointed or didn't believe me, they hid it very well. Last night we were all together at a restaurant eating tapas and it felt more relaxed and solid than it has done in years – as if a gap between truth and reality had closed and the floor had stopped moving away.

This morning my mother stood on her front step as I left, flanked by the perfectly shaped bay trees and emerging spring bulbs of her newer, smaller but still beautiful garden. I felt happy to be going home to mine but pleased that it wouldn't be long until I saw the three of them again. They are coming to visit in six weeks because Jared and I have planned a much-needed week away and they are going to look after the children while Liam builds the tree house in the field. I waved happily at her with the thought that we might be coming out of the decade when the relentless series of awful things almost obscured the good. My mother waved back and smiled too and I know she will have carried on waving until after I was out of sight, just as her own mother did whenever we drove away after Sunday lunch.

I turn into the plot where the children are waiting to welcome me back. They jump up and down, little fizzy sweets of feeling, and I am the same, undoing my seatbelt quickly, hugging them and noticing how good it feels to be home. As we walk into the house, I spot more snowdrops and the first of the purple anemones opening and let myself hope that this is the beginning of a more straightforward, easier time for our family of four – for Jared and me – and that this diagnosis is the closing chapter and clarity that I was waiting for in December. 2017 was a duckling-filled secret horror show, 2018 a slog of recovery that wasn't and 2019 the absolute worst. But 2020 – I try not to think it, but the thought makes a bud and then the sunshine opens the flower – 2020 is going to be the year I need it to be.

A week later and the phone rings. I let Jared get it. The past seven days have been an odd and exhausting mix: feeling almost high after the diagnosis and then an abrupt, balancing dip as I try and fail to reconcile myself with the person that the printed sheets of the test result describe. There's a dragging sensation from the bottom of my eye sockets and into the left side of my chest and the word for it might be grief. I have always been someone who notices, is observant, switched on and tuned in. But I am now also someone who missed 36 per cent of the clicks in the test. I have lost my understanding of who I am and of what the world around me is made. Might I have missed 36 per cent of my life and, if so, what happened in it? And what was I so focused on that I tuned out over a third of everything else?

I find the hyperactivity hard to accept too, though I have started to notice it: a constant need to move dressed up as being useful and the tiny but constant fidgets when I can't. I find I hate it, thinking of myself like this, moving 11,000 times in fifteen minutes even when I am trying to be still.

The news outside the plot has hardly been soothing. A dangerous virus has been keeping the people of Wuhan in China locked up in their flats and though — like so many things — it seemed like a distant disaster at first, it's coming closer. The stories from Italy are horrifying and Jared and I have a brief falling-out over it because I am becoming convinced that this is something to which we should start paying attention — and he thinks it isn't anything to do with us.

I sit with the tiredness of all this as I hear Jared answer the call, then come in and hand me the phone. It's my father. After a little small talk he tells me that he hasn't been feeling well, has had a fever and cough for a few days and was tested for Covid-19 this morning. The intangible, worrying international news story that's been tugging at my hem instantly becomes a tangible threat. In answer to my many questions he explains that he qualified for a test because he had recently been to northern Italy. I'm confused because he'd been in Switzerland just before I saw him, not Italy. He explains that they went across the border, over the mountains, for a day. A lunch, cable-car rides, smushed up against other people, touching all the things they had touched. We say goodbye and I hang up feeling steady and unflappable, a crisis is my calm place, but I am really worried about him and my mother too. Then I think of the tapas I shared with them, our fingers in the same bowls, and realise that if he does have this new disease then he could have given it to me and I could have given it to my family and they could have given it to so many others.

And just like that, the hoped-for conclusion becomes another chapter, and the next contender for the something I felt sure was coming reveals itself.

Monday evening in late March and I can see the annual daffodil show out of the living-room window and I focus on its cheery yellowness as we wait for the prime minister to say what we all know he is going to. I am still only partway through realising that, whatever he announces, nothing will be normal for me ever again. Mental illness is not the root of my problems – I can't be cured – and though there are many things that will help, I am going to be putting some of this extra effort in forever. With our first taste of self-isolation before my father's negative result, the almost hourly breaking news and the schools closing last week, I have been forced to process my diagnosis more slowly than I otherwise would. There is too much going on to hyperfocus on one thing. Now Boris Johnson's doughy face fills the screen and he confirms that, yes, we are going into lockdown. Lockdown – like the eggs I have just set in the incubator in the three days before hatch. Something was coming. That something has now come and even the most optimistic readings of the situation have us held here by it for months. And it is strangely okay, this dramatic and intense place – for now at least. This is my comfort zone, a coming into my own.

After the announcement we reassure the children who are still in an excited, summer-holiday bubble of no school and put them to bed with tales of all the things we are going to do with this wonderful time together. Then Jared and I sit in the kitchen with very large glasses of wine and talk about how we are going to deal with the time ahead. 'Do you remember when we thought last year was bad?' I laugh slightly hysterically, but I feel even and clear in a way I am beginning to understand. I am primed for sudden change and if I'm not flung from one thing to the next I will find a way to propel myself. My body chemistry has been set to fight or flight for so long that this is oddly comfortable. It's the day-to-day that makes me wobble and want to flee.

So, despite the genuine threat to many things and people in our lives, we talk in a united way that we haven't managed for so long. When I say that we should abandon any idea of not growing much and instead make this the moment we really go for it, he agrees and seems to mean it. The shortage of some products, the rationing in shops and the risk that now comes with going into the outside world have made the plot seem to him like the logical next step it's always been to me. Here we will still have the feeling of freedom, a little more control over our survival, the space for the kids to run wild and nature on tap. The sense that this was a mistake, a symptom, my symptom, my mistake has completely evaporated. We are here now, the law says we have to remain, and we are finding we are more than happy to do so. We smile at each other – the most certain thing each of us has in this newly uncertain time – and we plan our work on the plot, which finally feels like our solution instead of my problem.

I have jumped back in my garden today like a cool lake in summer, with Jared and the children at my side. We dismantle that fucking greenhouse with energy and joy, leaving it in a tangle of metal and a pile of panes that will have to stay in the drive until the tip reopens. It feels strange to look at the rectangle of empty space it left, and the trench where we'd buried the frame to stop it from blowing away. It is a rectangle of absence, arguments and disappointments, so I obliterate it with a sweaty hour-long frenzy of digging.

Now all four of us are busy remaking the veg plot, which will take us a few days. We are putting one of my long-held plans into action and replacing the hard-to-reach-the-middle square beds with long rows that will double the space for edible crops. The kids are helping rake the woodchip back while Jared brings over barrows full of slightly underdone compost and tips them where we have cleared. There is no need to weigh up whether we can afford to buy better compost, or whether I should wait until this batch is rotted down further because it's probably full of weed seeds – there is a delicious lack of choice. Another horrible constriction for many that I am privileged enough to be able to see the best of. Without options, decisions and the usual minute balancing acts of work/friendship/marriage/doing good/ staying sane, I am freed from the constant battles with myself. I have no choice but to go with it.

This new year's first heat came in a little later than its prede- cessor's, but I feel it on this early May day as I crouch to see a puff of forget-me-nots turning pink at their petal edges. Among their familiar softness are the occasional sentry-like tulips – few and far between because I haven't planted bulbs for two years but still beautiful. A sharp yellow, another white with magenta veining that makes me think of Vermeer and a silken purple-black that sets off the rich orange of the Icelandic poppies as if I had planned it that way. I didn't. But there have been chances to be glad of the things I did plan and prepare for here. In late March when my phone lit up with messages asking how to sow radishes or plant potatoes, I was pleased to be able to help, to have the seeds to sow myself when everywhere had run out and enough to give plenty away.

The loose white blooms on the scraggly delicate heads of Allium neapolitanum, white garlic, are beginning to open where I inexpertly planted their bulbs a couple of years ago. I've lost a lot of alliums to digging them up forgetfully, but there are enough left to enjoy. Their taller, more robust cousins, the purple Allium hollandicum, are busy shooting skywards in clumps; their straight stems contrasting with the soft mass of spring growth below. In the veg garden I've more than made up for our late start. The potatoes are in, hundreds of onions and shallots have been planted by the kids, and everything that can be sown has been sown, with more planned for the moment the weather allows.

In adjusting to the new reality of everyday life this spring – the worry, the risks, the logistics, the no school but still the work, the money worries worsening – I made some of my usual mistakes. Trying to overplan my way through the new landscape with a series of goals and expectations: a daily schedule, a radical budget and a pile of English and maths workbooks all bearing a large photo of Carol Vorderman's face. There was a lot to do and I did it all straightaway because it was an emergency and if I didn't I might forget. There were new forms to fill in to defer mortgage payments and apply for government help, a list of we-might-finally-get-round-to-this jobs to make, shopping to decontaminate, a dishwasher that never stopped and a house that needed cleaning twice a day. I tried to navigate this by scattergunning work pitches into the digital ether; joining village volunteer efforts and honking the car horn and clapping for NHS staff at 8 p.m. every Thursday evening. It's all falling away though, apart from our enthusiasm for the smallholding and an unreasonable lingering hatred of Carol Vorderman. Something gentle and more continuous is emerging instead and I like its flavour – more concentrated every day.

Later, while Jared works, the rest of us watch a documentary about the Hubble telescope that will do for homeschooling today. The roar of the space shuttle's rockets grips Sofya and Arthur but it's the thirty-one computerised visualisations of the 'pillars of creation' that have all of us gaping and pointing. They depict the cosmic wombs in which stars gestate, churning columns of apricot set in a teal sky. We sit together in a warm pile, our minds blown, bodies pleasantly heavy after days of manual work and too many late nights. I am half asleep when, just before the credits, a quote from Edwin Hubble, the American astronomer who gifted the telescope its name, appears on the screen. 'We do not know why we are born into the world, but we can try to find out what sort of a world it is – at least in its physical aspects.' I like these words a lot and want to let their meaning run around in my head but, as children's bedtimes are not for existential wondering, I take a moment to note the sentence down.

A couple of hours later, I am finally able to look Edwin Hubble up and follow his quest to find out about the kind of world he lived in. This calming meander through links leads me to another quote and I like it even more than the first. When asked about his beliefs, Hubble answered, 'The whole thing is so much bigger than I am, and I can't understand it, so I just trust myself to it; and forget about it.' If this genius – who would discover things about the universe that would change our understanding forever – didn't have a clue, then maybe it's okay that I don't either. Hubble tells me that I am lucky to be able to see it – the vastness – to feel it crackling above and to try to find my own way to connect with a little part of it. It isn't a threat if I trust it, if I trust myself.

I turn off the light and close my eyes and in sleep I am there almost instantly – the place my DNA knows. Alive with noticing – just as I should be: the path ahead, the changes in the undergrowth, the signs of water, a bird of prey in the sky to the west, a rustle – a predator? – in the distance. I see the herd as a whole, one huge undulating animal, and if a part of it slows or speeds up, threatening to break away from the mass, then I can catch it before it does. With a movement of my arm and body I keep the herd together, yet I see them individually when it's needed. A limp, the first telltale sign of a pregnancy, a tiny change in movement makes me zoom in on something small but newly important. I spot when the people walking with me are tired, sad, bored, hungry or fearful, often before they do, and so they feel safe even though this last grass we are chasing will soon run out. I will find some more though – it is my way, my job, my life.

However much I fear I am a madwoman, a crappy friend, a bad mother and rubbish wife, here in my dreams I know I would have made a bloody good nomad.

# ZOOM OUT

It is July, the tipping point of a summer that began in late March, as the world folded itself inwards, and is still going strong. I am on my way outside to water the mildew-prone courgettes with a well-water pump system I've fashioned. Elsewhere this month I hear there has been rain, but it hasn't been wet on the plot since winter when it was flooded. The Met Office says this is going to be the way of it now: a boom and bust of wet and dry and every year's heat pushing up a little further. An inconvenience for most of us for now, but already a matter of life and death for some. I'm working on a plan to store large quantities of rainwater in the winter so that I can give it back to the ground in the following summer's shortage. A drop in the ocean, a plaster on a compound fracture, but I tell myself that a drop is still wet and a plaster stops a little blood. I keep hold of these small things.

I drink my coffee and eat a banana so I can stomach the ADHD medication I was first told I couldn't have but, after another formal appeal by my saintly GP, I now take daily. It is not a miracle cure and I don't want to rely on it forever, but it helps for now. My mind is less scattered, more able to focus on a single task and then step away when it's time. The biggest surprise is the effect it must be having on my dopamine levels

– usually low in those with ADHD. After a few days of taking it I realised I felt a sort of calm contentment from time to time. Nothing dramatic, not happiness, but a quiet feeling of being still and fine that I've experienced before only in the mountains or on the back of a fast-galloping horse.

With my boots on and long sleeves between me and the dawn air, I step out. If I head to the top of the field I know I'll be able to see right across Victor's land – Footway Field – to Devil's Hole, the small hills in the distance where they used to make bricks out of the clay and, beyond them, to the wind turbines turning out at sea. There has never been a brighter sky over this land than this morning, but its clearness doesn't stop me from also seeing the thick mist moving in spirals and curls around the fence posts. As every day for the past four months, mist winds around my ankles like a cat, hovering just above the spikes of grass, the bird's-foot trefoil and clover in the field, obscuring the narrow path that the animals have worn to the gate and confusing the edges of the boundary trees.

Despite the months of dry weather, the sky is always full of water droplets: unexpected obstacles that get in the way of the light waves as they try to reach my eyes from the sun or the moon. As the wave meets the water it is forced to bend, scatter and split and then white becomes red, yellow, green and blue. To the layers of time and place I have already learned to see on this land, this year has added another: fogbows and coronae – eerie, shimmering patterns with extraordinary rings of soft colour that may or may not be more beautiful than rainbows.

Over the warm months these glimmers have helped me find one thing I had been looking for: gratitude. I have always known how lucky we were to be here and to be able to do this.

Accidents of timing and birth helping us buy our first flat a couple of months before the crash that meant people like us couldn't any more. I always knew to be grateful, but there was too much in the way of really feeling it. Now I do, at least in the moments when everything stills.

The makeshift pump is plugged into an extension lead that has been outside for weeks as if to tempt the rain. I turn it on and hear the water bubble up from the well, past the ferns that grow inside its brick collar, and then point my hose to the soil under the courgettes' leaves. I let Melissa the hen and her eight chicks out of their run and into the big field for the first time. They all stream out swiftly and I watch to check they stay together. As I move through this morning routine I am searching for signs – of the season, of the past – as I now always do. There is a cacophony of voices and sounds to tune in to. Victor's old-fashioned phone already ringing, geese honking, a faraway tractor, the scratch of a hundred-year-old hoe, Harry's pigs squealing, Susannah's steps on the drovers' path, Joane's slower pace to the coops and the hum of the battery chicken farm down the road.

Under all this I hear a baby. I concentrate and help these cries of many yesterdays become a roundness of skin and milk dribble. The child is tied to his mother's back and though she bends carefully so as not to tip him out, she ignores the wails for a while so that she can finish clearing the ivy that's finally loose enough to pull free. It's her I am interested in – it's always her.

When she's stripped the dry vines back, she starts moving rocks. One is far too heavy, made more cumbersome still by the squat position she must adopt to protect her little son. She lifts it anyway. Stubborn. Relentless. And then finally sits on a stump next to a vast and newly felled trunk. She looks at her blistered palms for a moment, eyes flicking to where an axe is leaning against all that is left of Andredesleage's Mother Tree.

The woman moves and I think she'll take up this destructive tool again but instead she reaches behind, unties a knot and loosens the shawl that has been holding the baby. Her face changes as she brings him around to her: hard becoming … not soft but flexible as they smile at each other. He latches on, little sucks at first and then tiny pin-like fizzes of milk let down in her breast and he gulps then swallows. I feel it too, prickles of milk under my areolae.

The child settles and she wipes her sweaty face, leaving a smear of mud. Dirt from a place she has claimed and will care for. In this small spot she's the primary link in a new human chain that runs up to and beyond me. She is the first to strip this place back and turn the earth into survival and shelter. She is the one to credit. She is the one to blame.

This woman is the settler: the invader, the oppressor, the destroyer of all things.

She is the guardian: the solution, the protector, the bringer-to-life.

The peasant. The queen.

I search for her shadow. I walk in her footsteps. I tremble in her wake.

I kiss her ankles. I tread on her toes.

I have been at my desk this morning but now it's lunchtime and we eat together under the orchard trees. Sofya is sitting opposite me, eleven at her next birthday and busy spooning food on to her plate (and the table) while telling us about the fully equipped equestrian centre she has been running. It is an impressive affair complete with rodeo lessons, a horse called Nightmare and a dressage arena where her brother, who has little idea what dressage is, trots round on an invisible pony. Arthur, who sits next to me, will be seven soon: fringe flopping over his eyes, yesterday's mud on his cheek and knees, hay stuck to the back of a baked-bean-stained jumper. Our new ducklings and gosling are standing as close as they can to him despite the fence, because he is their favourite person and the only one who can persuade them into their house at night. I stay close to him too, this boy who looks like love and smells like a farmyard.

Everything I am serving myself now was grown here. It turns out that it is possible for me to have a really productive year on the smallholding: I just need to put in thirty-five hours a week and have three dedicated helpers. I remembered to do second and third sowings for a continuous supply of vegetables, enough to share as well as for my family. We staked almost everything before the wind could snap it and had the brassicas under nets months before the cabbage whites emerged. The vegetables compete for space with the self-sown larkspur, sunflowers, poppies and calendula that I spotted in December's mud, and there's a new cutting patch with zinnia, cosmos, chrysanthemum Hippy Love Child, mammoth dill and foxgloves that we cut down once and are now coming back for round two. The meadow is tall with paths that I marked out with sticks and Jared cut with the mower. Arthur likes to run his remote-control car along them and the cats are convinced that they are feline tree-climbing launch pads. After lunch we're going to walk in it and sit down for the briefest of moments to be eye level with the wildflowers among the grasses, and see the world of insects making their lives within.

I take a forkful of cauliflower and it tastes nutty and good. It has taken me four years of sowing cauliflower seeds to get to this point of actually eating one. Some things do take more than one season and one pair of hands. Jared and I look up from our lunch and nod at each other, acknowledging this tiny, delicious and significant achievement. There will be no time to rest on our laurels; there is work to be done, plenty of mess to tackle and we are still screwing stuff up. Yet it doesn't feel quite so perilous. The previous three seasons of mistakes and disasters feel like hardcore training for the time we now find ourselves in. And the I, I, I aloneness of last year is slowly becoming us; we.

The children have finally gone to bed and I am in the bath, living dangerously by reading on my phone. 'We are entering an era of pandemics – it will end only when we protect the rainforest,' reads the headline of an article I probably shouldn't be absorbing myself in at the close of this long day. An era of pandemics. Great. I read on, facts I know, some I don't and new connections forming. Powerful people, people who look a bit like me, have been snatching from the rainforests for a long time – from other land too – and often from those with a deeper understanding of the ways our lives are woven with it. These thefts are catching up with us now and the consequences would serve us right were it not for the predictable horror that indigenous peoples are hit hardest.

These scientists say that this pandemic – and those to come – reached us humans because of our exploitation of the natural world. This ecological smash and grab is why I have to worry about my parents going to the supermarket, why grandmothers from Basildon haven't held their new grandchildren and why Ismail Mohamed Abdulwahab, a thirteen-year-old boy, died alone in a London hospital. Headteachers are delivering sandwiches to hungry children missing their school dinners because we have been crashing around in the wildest places and taking from them – taking more than we bargained for.

I put my phone out of splash range and wash my hair with as little shampoo as possible because we'll bail the bath out with the mop bucket later and give this water to the plants. My ears fill as I submerge my head, full of thoughts of lands vandalised. I can't absolve myself from this rampage – I've felt its benefits too and ignored the reality of damage and destruction of places and people.

It is all connected. The maps and the lines, the thoughts joining this to that and leaving me in a spin are not madness. They are sanity. Everything is interlinked, everything is important. The rainforests are burning further and faster than they did in last year's record-breaking fires. They have been alight since the colonialists showed up and they – we – started extracting everything. The djendjenkumaka tree still stands in the middle of the forest, for now. It shelters birds, mammals, insects, the underwood and the Lokono and Saamaccan people under its ancient canopy. But flames are approaching from one direction, the devastation of the virus from the other, and I don't know how long it's got left.

I sit up, water running down my face into the bath and later back to the ground. If we are all connected – people, places, animals, trees – if I and my actions are as linked to charred stumps and mutating viruses as I feel they might be, then is it okay to pace myself, to have a car, to turn the lights on, to laugh? I am not sure about the car (though we still have one) or the lights (until they aren't at the expense of something else) but it has to be okay to laugh even when this vast network of connections becomes visible. There must be a way to crack a smile without stuffing my fingers in my ears or shouting 'la la la!' to drown out reality – I just haven't quite worked it out yet. None of this is easy and these kinds of thoughts still make my words come too quickly. Switching off is hard too, and the conflict between this need to stop and to go has me pinching my side again now. Another bruise.

I am so new to this. Lurching around in adulthood; a beginner on this smallholding; an intruder on the land. I'll be an interloper even if I live here for the rest of my life. It must take generations for a settler truly to become a custodian and I want to find out how it's done. The Gaandlee Guu Jaalang – the Daughters of the Rivers of the Haida First Nation people – are who I would ask for advice if I had any right to. These Haida matriarchs have filled my newsfeed and thoughts since they announced their occupation of two ancient villages, Kung and Skaaws, last week.

The Haida people are the opposite of newcomers but even they must have arrived as settlers long ago, making their home on an archipelago dotted off the North Pacific coast of Canada. Since then they have grown through the land like a vein of mineral and now they are upholding their laws to stop a luxury fishing lodge reopening and endangering their community in the midst of a global lockdown.

I want to ask the Daughters of the Rivers how they learned to be these guardians of a community made up of more than people, but they are busy with more important things. So, I dry myself and sit on the bed re-reading their recent statement to the press instead.

The Haida people are sick of the fishing lodge's wealthy American clientele trespassing and spoiling the earth by using it as their playground. They have protested hard about many things over the years, been arrested and suffered more than I can comprehend to keep their link to the land. All but wiped out by the smallpox, the Gaandlee Guu Jaalang know and fear the impact this new pandemic could have. They are staying put, invoking their kuniisii, their ancestors, and say they plan to survive 'at all costs'.

I click from their statement to a video and the Daughters of the Rivers, the mothers of their people, fill my little screen. 'We are here,' they say. 'We are here and we never left.' A message to the media, the lodge-owners, the government, and I wonder if they were reminding themselves at the same time. The Gaandlee Guu Jaalang look straight into the camera lens and stand behind their spokesperson, Kuun Jaadas, as she explains how it is going to work now. How it should always have worked: 'As people of Haida Gwaii we uphold our responsibility as stewards of the air, land and sea and assert our inherent right to safety and food security in our unceded lands and waters.'

The Haida women pitch blue tents under evergreen trees. 'We are here. We never left.' They take to their fishing boats and sail the waters that lap the edges of their ancestral shores. Grey sky meets grey water and they glide across its surface. On the shore the Haida people sing to their leaders, a melody made of gratitude, drums, voices, tears and the swish of legs wading out into water to be closer. 'We are here. We never left.'

'I am here,' I say, trying it on for size, but it doesn't fit; I haven't learned how to be here yet. I can't promise we will never leave, however much I want to stay. However much I want to live in a world where the Daughters of the Rivers are in charge of everything, they are not. And that's why I will keep on trying and why more of us are going to have to learn – and quickly – what it means to be a steward of the air, land and sea.

I put my phone down to get dressed, looking out of the window into the evening. The goats are not yet in bed so I go out to do these last jobs, a routine that is only a few years in the making. The ground under me as I open the gate sits at the crossroads of so many urgent issues and the light of the end of this midsummer day makes them very visible. 'Hands, face, space,' reads the government's snappy new pandemic slogan. These emergency orders presume a safe home, having space, a hot tap and an outdoors where you are welcome. We have enormous privilege here and it has never been more obvious that it's not a happy accident that we have this and others do not. The Black Lives Matter protests following George Floyd's murder have come at a time when people of colour in the UK and across the world are at greater risk of dying from this new disease and more likely to work in jobs where they will be exposed to it. It is not lockdown for everyone. And lockdown for most is not the same as it is for us here.

Here. We are here aren't we? Even if it will take generations to sink in.

262

A few days have passed. The children are watching a film and I am at my desk writing again: writing my own ending, the end of this book. I've read it back, this story that has kept shifting – on the page, on the plot and beyond – and still refuses to be pinned down, even now when I really need to finish it. There are too many ideas, leaps, far too many 'ands', 'buts' and long sentences. On one reading it seemed as if it wasn't nearly enough to go out into a world that is suffering and struggling. I read it again and then it felt like far too much for anyone to deal with. Trying to do all, trying to be both, joining everything together without pause – and, but, and, but – and leaving gaps that others would have filled.

Like almost everything I do, I have made writing this story difficult and complicated. Writing is difficult though; this story is difficult and life is complicated and contradictory. It has taken courage and stubbornness not to tidy it all up, to resist the urge to make the narrative more presentable by editing myself out. And (another 'and') perhaps it won't make sense to anyone else. But (another 'but') at least it will be true.

My back is stiff from the hours at my desk and so I take a break and go outside, kissing my three favourite people on the tops of their heads on the way. I go to the hollow tree to see if anyone has laid an egg in there today. One of its huge branches has come down in strong winds and the goats have set about it delightedly, balancing on their hind legs to reach the leaves of their favourite meal. This tree that they perform acrobatics of desire to eat is Salix caprea: goat willow. I knew I had identified its species when I found an image of a goat stretched up on two legs to get at the branches. It could have been a photograph from my plot but was instead an illustration from *Herbal*, the 500-year-old book by botanist Hieronymus Bock.

Amber and Belle are at work stripping the delicious bark surrounded by generations of livestock who have set their hooves against a trunk like this. As I look on, the trees fill themselves in densely around us all, the sky is slowly obscured by leaves and the air cools in the new shade they cast. I hear a thud, followed by a moment of absolute silence. The goats are motionless, their eyes wide, and even the nearby hens stop their scratch-peck for a second. Another sound, louder now, breaks the pause: a faraway groan and a crash. In its wake we are stilled again – both by the fall of another great oak and the loss that each of us knows through its absence.

Then, within an exhale, every living thing moves on and turns to the light, celebrating the new warmth. The trunks thin out again, the sun reappears and I remember how much I love this clearing in which I have made my life. I need this space, light and heat to exist – to grow. Progress is necessary, change comes anyway and I am grateful for it. Yet however deeply I know the present is true, the outline of the past persists – its leaves always in bud – reminding me of something else I love and long for, insisting that this is a place where the ancient forest should be.

A forest or a clearing: which is it that I want?

My break from work is over but I haven't returned to writing quite yet. Instead I've pulled up the 1939 Register of names and addresses where last year I looked for our smallholding and failed to find it. With the archives closed by Covid the question of why this place was built and who lived here first has remained unanswered. I had convinced myself that the story could end without this missing piece, but this afternoon that feels like a lie. For what I promise myself is the last time, I look at the words on the screen and soon see that our plot has not magically appeared. I'm about to click away when one of the house names sparks something in me. It's similar to ours but more fitting and yet I haven't noticed it before.

Without house numbers there is no easy way to tell which name corresponds to which contemporary property, so I flick through my folder of maps and records and try to make sense of this possible clue. I find myself reading a document that I'd previously skimmed and dismissed. This time, I slow down and look carefully. And there, at the top of the second page, is exactly what I have been looking for. Our plot did go by a different name in the past – one that is an exact match for the title I've just spotted on the register.

I have found it: a single word that leads me to a widower called James Sanders who occupied this place at the outbreak of the Second World War. He was a small-time poultry farmer, living off this land, as I had hoped he would be. A little more research leads me to Lucy Kempson, a widow of seventy-seven. She used to run a grocer's shop in another part of Kent but by 1939 she was living here as 'unpaid domestic help'. I raise an eyebrow and share a wry smile with her inky imprint.

This plot's real name is Oakleigh. I love how it sounds when I say it aloud, the word breaking down into its component parts in my mouth: Oak Leigh. Leigh. A reflex. I look it up. Leigh: from the Middle English 'legh' – an open ground. The word's thread runs back to the Old English 'leage': a clearing in the forest. I find it in the languages before: 'lauhaz': meadow; 'lówkos': field; and then, there it is, in the Old Saxon and the Middle Dutch: 'lōh' – a forest. The forest has been here all along. Oakleigh. Oakleage, Andredesleage. A forest and a clearing: it can be both and so can I. I can be a forest and a clearing too.

The rain begins to fall at last as I start to type into this final chapter. I know that two names on a screen don't turn this into a proper ending; that the threads aren't tucked in and nothing is simple or fully explained. But this is not the end. Nothing was over in December because a decade was done; it wasn't the end in February, as a doctor wrote an acronym on a piece of paper; the story wasn't finished when the pandemic started and it is not ending now, even though I have finally harvested ten perfect broccoli. I don't even have room for them in the fridge.

The end might not exist and if it does it is somewhere over there, in the then. And I am here in the middle, the place where everything just is – the comma and not the full stop. This quieter space between the question and answer is where I need to live and it is where the story lives too.

I put off the chapter's last sentence for a moment longer – not quite ready to commit to the final words. My emails distract me for a moment. 'During this apocalyptic time . . .' begins one that goes on to try and sell me loungewear. I snort, imagining cashmere jogging bottoms as the only logical response to the end of the world. Maybe they are?

'Apocalyptic' again, though it seems like less of an exaggeration today than it did in February last year when the moors burned. This strange, difficult, frightening time (we won't know what adjective to use for a while) could be the start of an apocalypse – part of a collapse that changes everything for us and for the earth. Yet it could be something else too: less final, more complex – still awful, but with a better path out than the one we took in.

Either way, it is not an ending. 'Apocalypse' hasn't ever meant that.

Apocalypse:

old English
via old French;
from the Latin – from the pulpit –
from John of Patmos,
and, since I was seven, from the dictionary
as:
the end of days.

Apocalypse: it's the end of the world that we do not want –
though you really wouldn't know it
from the cut of our jibs,
and the thud of our boots.

Apo
calypse: a collapse,
a cataclysm,
and the screaming – always with the screaming –
fire against skin.

Apocalypse: from 'apocalypsis'
– a four-beat hidey-hole of a word, which
jumped from nave to knave a thousand years ago.

Full of visions,
wild, wide apparitions and
stream-clear insights;
new ways to look at old things.

'Apo', Greek – un, off:
*Take it off, let it go, get it away from me!*
And Kalyptein:
*Hide it from them!* (An island, a forest, a map.)
*Cover it, keep it safe!*
Conceal.

Apo kalyptein: 'apokalupsis' – an uncovering,
an unearthing;
a little round something that has just been dug up.

Apocalypse:
an ending beginning in 'kel':
to hide, to save, to keep this
root word. This PIE word.
A Proto-Indo-European word,
saved from this language of starts,
of myths and
lost,
until we found it –
down the back of the sofa?
No.
We guessed it back, found branches
and made an estimation.
We made it up.
(*It's all made up.*)

Kel, kalyptein, apocalypse.
Kel, kalyptos, eucalyptus:
an evergreen tree
whose leaves are crushed
for aching joints and open wounds.

Eucalpytus:
hard wood of tarnuks –
water bowls –
carved by the Kulin
who left their marks on the falling water's edge.
Bark-peel scars and
knife-written notes
from those who paddled their narrow hulls.

Hull: back to hulu, hulla, huls and then –
yes –
we boomerang to 'kel' again.
A linguistic seance.
Whispers along the river, binding roots and
red
bark
hulls.

Hull: the concealing case of a tree
(eucalyptus)
named for this: its hull, its operculum,
its elf-hat cover slipping slowly from the bud.

Operculum:
the little lid that protects a flower,

that covers the cortex – the soft inside our skulls
where consciousness rests –
though it never rests.
Grey matter, dark matter
chattering and chattering,
while we wait for the end
that might be here.
That might never come.

That we dream of at night.

Night: neht (Old English).
Light: leht (Old English).

Leht –
lightning:
old,
new,
electrostatic –
under my lid,
forking out of the cloud cap above
and searching for

the path of least resistance.
A current of bones,
the smell of burnt hair
and then:
toast.

Unless?
Yes.

Lightning –
leht: the first ray of a new day
on my lids as I wake
warm from these hallucinations.

Light: from the dawn
on the path;
light from the candle
on the wood, on the hull, on the walls of the hold of
this small ship
that is still empty –
stuffed full –
an apocalypse of stems.

Light-
ening the load.
Leht-
ening this weight.
Lightning across the sky again and again and again

yet

I   am   grounded
And moved on
(gently)
by the power of the current
of the vessel
of expansion
of intention and of
my grey cockerel crowing in the plot's eastmost corner:
his song of what it means to be earthed.

# EPILOGUE

Evening light: my favourite. It hurts to move my face towards the open window to see the sun on the top of the old oaks' canopy, but I do it anyway. A square of two-tone trees, skyward leaves lit up as if from their insides. Midribs, their main veins, give out a clairvoyant glow of future-fire. I know that this ordinary, magical thing will happen again tomorrow and the next day, but I won't be here to see it – not from this vantage point anyway. This is my very last time.

My little oaks don't yet stretch up to join in this end-of-day dialogue with the sky. But I see them over there; a little to the right, behind the ripeness of the veg plots. Swollen knuckles, permanently locking my fingers around an invisible handle, called time on my own sowing and growing a while ago: a new layer of sadness and frustration that showed me how to step back, pass on and share. I still have peppers in summer and they taste the same as the ones I used to grow; even sweeter perhaps. Fluid-filled vacuoles, hunkered down in the honeycomb of each fruit's cellulose skeleton, responding to the light, the air, the stars, with a flavour that can never quite be repeated. Each bite is a suggestion of what I've lost and gained, the weather, the soil's hidden life and the hopes and struggles of the other hands who have tended to them. The veg garden is now a blur that I can decipher only with my tongue, but the little oaks are still clear to me. They've come a long way from the sticks with roots that we planted one rainy March afternoon: healthy, green and over ten metres tall now. I suppose I should stop calling them the 'little' oaks – but it's an old habit and old habits die hard. Huh. I hope old habits and old women can die gently too: it sounds less painful that way. Though, if I'm honest, it sounds a little boring as well and I'm not good at boring. I've never done gently well either. The hard way it is then. Bring it on.

I hold my eyelids apart until the sky dims and all the trees revert to green, and then I let them fall and enjoy the blackness. It seems I am dimming too: going back to the land – a place I used to think I'd already returned to – with a carful of stinking cat and a fistful of leek ends. Fuck that fucking leek! So arrogant to believe I was ready to go back to the land that day. I couldn't return to a place I had only grazed the surface of before; a place I didn't know or understand or even know where the understanding might begin. I am doing it for real though this time: going back to the land, if it will have me.

This is the moment when the seed I sowed way-back-when flowers for the first time. Of course it would turn out to take this long to understand what I have been trying to all along – that the end result was the journey: all of it; the whole shebang; every breath, every touch, every fall and every glimpse of something good, bad, ugly or majestic. Understanding was never the point, finding the answers to all my questions was never going to be possible. Walking along, asking, wondering and finding the right place to slot into the vastness (without trampling too many soft things along the way) was everything. And it was so beautiful to try to do it, and fail, and then to try again and again. I'm smiling at my mistakes and it hurts my jaw to do it; mandible and maxilla alliterations of pain that turn an easy smile into something more: better, deeper. That is usually the way with pain.

The air has the taint of bonfire and I close my eyes and breathe it in. It is the smell of a chick in the pocket of my dungarees, kissing Jared against the wall of the goat house on a hot August afternoon, crying alone in the shed and working it out with the turn of a sharp-edged spade and a barrowful of woodchip, sinking myself into this place, learning how to make it count. Freckles and fires with the little ones, who became the so-much-bigger ones in such a long/short time. Rough hands and rough days too. The awful things, the things that fixed me to the spot for a while and the slow limp away from them with new lines and grey hairs. Red-rimmed eyes more alive every day to how beautiful the pattern of holes was in the orange dahlia's petals; even though it was made by small slugs on their way to ruin my strawberries. Those slimy, sluggy bastards.

Bettering, worsening, perfecting, destroying: it all depends on the way you squint at it, doesn't it? A consequence of little holes, small spaces that opened up to be filled with love and sweat and all the other things that brushed against my skin here on this plot.

It is thinner now: my skin. The moth's wings will turn to powder if I try to help it find the open crack in the window it's beating against. I've learned to let it be; to do only what I can, wherever and whenever I can do it. Oh, I did a long stint out there beyond the hedges, but what it's like today, what has happened in that expanse of roads and towns and fields, seas and continents escapes me now. It has all disappeared and this place has become all the continents I need. Sometimes, on a quiet evening as the serotine bats flit to a beat I will never hear, I find all of the globe in the air they move through. A world of holding my children on the daisy-peppered lawn while we listen for nightingales. Nightingales? I might have that wrong. The white storks though, they were real. Nesting at the top of a tree – an oak, of course – not that far from here. Pale miracles with chicks that hatched, fledged and flew off in defiance of the six-century gap between their last flight and this new journey in the summer when everything stopped. Did they come back and nest there again the following year, the gap becoming a blip? I can't remember: it's all so long ago and none of it has happened yet.

I sleep a little and wake to the day's last light and a moss carder bee passing the window. It lands on the rose that I propagated a few years back; a quick rest on its way to the clover it loves so much. I gave this rose its name: 'Sunrise'. The name of the rose. *The Name of the Rose* – that rings a bell somewhere, an echo of standing at the window by my father's bookcase; hands on a creased spine, heart away in the valley and eyes looking out over the garden of my childhood and catching a red-haired boy in the distance as he disappears into the trees to capture tigers. Every story tells a story that has already been told. Does it? No. The Sunrise Rose is here only because of me. Because I did and I could and I will and I would, in the way that only I can. But also, yes, every story is a repeat that repeats itself over and over again. I cut the stem that made the Sunrise Rose from a bush that came, in its turn, from another – one I propagated decades before from snips my mother gifted me from her own rose garden. A thread made of loops: women, roses, gardens, stories, breaths.

The bee lands on the petals of this, the most recent link in a chain that stretches to here from a long-gone garden. My mother's garden is now stones and thigh-high grasses. The broken walls and weeds with meadow flowers poking untidily above the rubble would drive her absolutely mad. I'm surprised that their very existence hasn't summoned a ghostly frenzy of chopping, digging and neatening. But, on the plus side, she'd be thrilled that next door's trees – the ones she hated in a way that no one, before or since, has ever hated a tree – blew down one stormy day. Their trunks still lie somewhere under the brambles, rotting gently and sheltering another world of tiny lives. My mother's garden is now a buzzing, teeming, wild place of hidden passageways for the voles and mice who dedicate themselves to eating the nearby plots' pea seeds before they can germinate. On balance I think she would have loved to see this expanse of mice, flowers, birds and know that we played our part in making an equal space for everyone – even in the centre of the big city where she gave birth to me. A city remade from the best of the past with the best of the present: the only possible future and a world we had to rebuild even though it was already far too late.

Or maybe this is nothing but a fairy story I like to tell myself: a hopeful fiction laid over a ruin. Because surely, when the dial was turned up another degree and everything they had said would happen really did happen – and worse – it was chaos. Everyone panicking and running around screaming, waving their hands in the air pretending to be shocked – as though this was something being done to them, instead of done by them. And if that's the case then my mother's garden is long gone still, but under concrete and breeze blocks and the mechanised quiet of slow-choking air. It's now a place that rarely feels the energy of a life pass through it. A place that isn't a home to anything, because there's no need to make homes for things that no longer exist.

I don't know which of these stories is true. I used to and I know I played my part, but I can't catch it now: what happened out there, what's happening today, no matter how hard I try. Truth isn't a solid thing anyway. It's not like one of the muddy rocks that

Arthur liked to dig up and clean, turning the washing-up-bowl water brown as he exposed their sharply defined edges: the clarity of quartz. There are always truths in the lies and lies in the truths, so what happened is academic. I'll let it go and look up at the darkening sky of now instead. Never a truer thing existed than this sky and there never was a sight more made of lies and tricks of the light. But none of that matters when it glitters above me like this: another map, another thing I won't see for much longer.

Drawing the outside air into my warm body, in the pointless pretence that they are two separate things, has become an irritant and there are only a few more times that I'm prepared to do it. So there! I'm going to be a difficult person to the end: always walking the opposite way with my ears, eyes and heart open, making everything harder than it needs to be. This is who I am and what I do and, finally, I don't pretend to be anything else. I make no apology.

I feel full of everything and completely spent. The casing of a squash seed on the soil's surface; a husk from last year's fruit that has already made next year's harvest happen. Have they felt like this too, the women who died here before me, in this place under the oaks against the sound of trees in the wind pretending to be running water? Wild orchids in the meadow. A dragonfly tipping down towards the pond: hovering and slowing as it prepares to get closer to the water's surface. And then a swift dart: down, up, back. Back. Back to where we have all been trying to get to. Back to the beginning. The heart of it all, the earth's core. The wild ride through the kernel of its secrets. The incredible heat and churning of molten lava with its promise of a cool future of mountains and moss-covered rocks.

I am seconds away from it and I still don't believe in it. Of course not. This end is another fiction: a deluxe death, tied up neatly with a ribbon and a bow, not the real one I should expect given what we have all seen. I should have an anonymous end: wires and too-bright lights with no one who loves me nearby. Or a desperate one: clinging to a tree as the waters rise full of bodies that explain, in a series of bloats, that I am next. This simple kindness of death, on a plot that I have watched over and has watched over me, a place that I leave stuffed with more life not less, on a land I will never fully know but am now part of, must be a fantasy. Because surely, by now, the plot, the oaks, the moss carder bee, the rose, the love: none of them exist. This death I want so much just can't be.

But, stubborn as I am, I am going to make it be if it kills me. Ha! My head-forward gait, an inherited, determined movement full of purpose, comes in handy right now. Bloody-minded to my last breath, I take a run at the death I want and I grab it and bite into it as though it's the last apple in the orchard. I close my eyes and will it to be this way, letting the last air out of my chest in a relief of long rattle: the juddering click-clack of the crack willow's branches on a squally day. No in-breath comes. I wait. Still nothing.

Nothing.

And then . . . then . . .

Can there be a 'then' after I've emptied my lungs? The final exhale is a full stop rather than a comma, isn't it?

I wait

and wait

and wait,

ah!

I wait a little longer, and soon enough it does come: the 'then'.

It is quiet and I am truly still for the very first time. So this is what still feels like: sure and not set apart. Going, going, gone . . . but also more here than I have ever been. I don't need the air any more because I am of the air. I don't need the earth because I am of the earth and the wind and the trees and the rain. And I am of the tame robin too, another Robbo, as he lands on the windowsill and, knowing death when he sees it, hops away again without a care. Off you go, fat little thing! I am of the badger-faced hedgehog, snuffling unconcernedly along her path, the ladybird moving towards the blackfly on the lupins and the tawny owl getting ready to hoot and hunt. We all come and soon go from this place, leaving prints behind to be brushed away or followed.

Stillness gives way to movement: I am time-travelling again. Flying off the horse's back through the valley's air and landing face down in the Runnel; the girl with two left feet on her knees in the playground, on the woodland floor, on the cobbles outside the pub; a body turning itself inside out ('This is awful . . . tell me I can do it. I need you to tell me I can do it.') and a baby is born on the bathroom floor ('It was you all along!'); making plans side by side in the sheepy idyll of a Welsh summer; sowing seeds in shortening autumn evenings; milking goats and wild goose chases; watching the woodpecker fly over the bean poles; crying, throwing, smashing, apologising, kissing; pumping water up from the well and discovering that the can is becoming too heavy to hold. I stretch myself further: making a clearing in the great forest; holding a candle up to eggs I take from my apron – still warm from the broody – and seeing the spiders of red veins in them; closing the gate behind me after letting the cows into Church Hams pastures and then, moving over the centuries in milliseconds, bashing fence posts in to divide the land and writing 'Oakleigh' on a makeshift sign. I am everywhere all at once: whooshing, zipping and somersaulting along the thread that I felt sure existed, but had never been able to see in full until this moment. I am part of it now. I am a current running through every fibre and reaching out to every living thing as I do. And yes, just as I thought, this thread isn't on its own. It weaves over and under so many others, an infinity of stories touching in more ways than I could ever count, that together make a vast, eternal, universal rope with the currents of life and death, finales and beginnings, running through and between every thread. I see the sparks as some start to unravel, untether, unspool, unmoor; their fraying ends trailing away into a black vacuum, while others weave themselves back into the rope, happy to be home.

Another flash in these charged fibres makes me notice that the rope is wound around something. I reach out because this must be it: the spool, the reel, the spindle, the mooring, the thing I have been searching for. The family home, the final destination that I felt myself coming away from and knew the hurt of separation like never before. Sometimes I looked for it over my shoulder as if it lived only in a forgotten past. Then I tried staring resolutely ahead, attempting to lead myself back with imagination and big plans. Well, at last I am near and I plan to hold on to it forever, this solid heart of everything.

It takes supreme effort to connect with it – the central cylinder – and, when I do, I feel something hard with deep grooves in its surface. This huge and solid object is not rock, not polished nor static; it is something permeable, alive, damp even. This thing under my palms has the feel and smell of bark and suddenly I know what it is. I wrap myself around the trunk of the huge oak tree and, for a half second, I hold on and know her and am her: the rings, the sap, the roots, the mother. Then there's a jolt and an awful lurching hook as I am ripped away. Nothing under my palms now; nothing anywhere but colour and wind that I swirl through as I fall and fall wondering how the hell I am still that child, plaits undone, plasters on my knees? Why did I get falling as my thing? I could have had flying, leaping or X-ray vision for god's sake. I would have taken being able to touch my nose with my tongue over this – over always being the one falling down.

Then, as abruptly as it started, the freefall ends with a surprisingly gentle landing – like sitting down on grass. I am not hurt and am quite still again, but it doesn't feel like the complete stillness of death. More of a living kind of feeling – not quite, but nearly – living-adjacent at least. What the hell is going on? Have I managed to screw up dying? Maybe I wasn't supposed to touch the centre of everything? Over-reaching is my speciality after all.

Think, Rebecca! Wherever I am it is quiet but not silent. I don't see anything yet but I can hear birds, a constant electric hum in the distance, an occasional engine, the screaming of pigs, a faraway voice or two and even an old phone ringing. There's a familiar smell in this air. It is rich, sharp and fresh – ancient and new – and the way the dappled light falls on my face and against my closed eyes can only mean one thing. I remember how to open my eyelids and, though it takes a moment to adjust to the light, I know I am back on the plot before I see it. When my pupils have expanded enough I find I'm sitting under the oldest oak tree and it all looks the same but a different kind of same; like standing on a hill on the other side of the river and seeing everything you know so well but with the light now falling on it from the opposite direction. There are leaves above my head and a cold, blue sky too. Though the sun is shining, the air is refusing to warm in response to it and the ground is hard with a frost that hasn't melted yet. The pond is frozen solid for sure; I don't need to see it to know. These kinds of days are my bread and butter: beautiful and biting.

I'm still blinking and wondering when I hear an engine in the distance getting closer and closer. There's a blue flash behind the hedge and then the car, stuffed full of lives and objects, crunches onto the driveway. I'm not sure if I still have breath to hold but I am holding on to something as the car parks and I wait for what I know is about to happen but cannot be happening. A lifetime of seconds later and, as I knew he would, the man opens the passenger door and steps out into the January day. He has his back to me, as he often does, and I stare at it for a few seconds, overwhelmed, before walking quickly towards him. I stand very close, pressing my own back against his – a trick I learned a while ago, though maybe not quickly enough. Each hill of my vertebra fits into each valley of his and makes us two halves of a whole. I feel his muscles loosen a little and he leans back into me, unknowingly of course. He doesn't turn around and, though we are so connected in this different kind of way, I find it still hurts to meet the deliberate wall of his back. I wish

I could remember all the things. I wish I could remember if he does eventually turn to me and if I ever really see his face.

The rear door opens and I move back. The little girl can't contain herself a second longer. She jumps out and runs around the car in a blur of six-year-old, a single curl helterskeltering in her otherwise straight hair. She opens the other rear door, reaches in and helps the littler boy out: a sleepy, grumpy wobble in his step and a plump fist in his big sister's hand. His feet are hidden by brown boots but I know the upward curve of their tops by heart: fleshy, puddingy little toddler feet that demand kisses, no matter they've been dancing shoeless in the mud.

The woman opens her door last and I know how she is feeling as she straightens up and looks out across the plot for the first time, imagining their new life stretching out over it: rushing excitement, the beginnings of an out-of-control web of ideas, plans and connections spreading in her head. 'I can't believe that we live here, that this is all ours!' she says to the man and I bristle. Reflexively I find myself shouting as loud as I can, my voice ricocheting silently off the old garage, 'It isn't yours and it never will be!' Nothing happens because (damn it!) she can't hear me; of course she can't. But before I get too frustrated I see them: the sound waves suspended in the air waiting for their vibrations to be provoked into action. And as she breathes in, yes, she sets them off. The woman catches something – a half-whisper – and it's enough for a first flicker of recognition to appear in her eyes. I realise that I will find a thousand other ways to show her that this land isn't hers. That it's not a gift, that she didn't earn it and it's not a prize that she won. I am going to prepare a lifetime's worth of sharp lessons until she learns that this land is on loan: a job, a responsibility, a promise to repair that's all part of a full, hard and wonderful life of letting go of everything she thinks she knows. I will show her how to be a steward rather than an owner, because this place is not something anyone can own, has ever owned, whatever they've signed or paid and whatever they've been told.

I see her clearly in this cold light, every cell of her: so fragile and unaware. She turns to him; reaching out, pointing, speaking in excited twirls and spirals and explaining how everything will be wonderful from now on.

And it will be.

And it won't.

She is taking everything in all at once and giving it out, in little bite-sized chunks, to the children who are largely ignoring her, little missiles of excitement that they are. She points out the berries on the holly, wonders about the blue plastic pipe (keep it, it will be useful), worries at the sight of the piles of leaves (keep them, they are useful too) – and yet she sees so little. She hasn't spotted how, as she moves towards the man, he makes an involuntary, almost motionless, half-degree turn away from her. She doesn't see that with each loud and over-the-top word she fires into the space between them, a thoughtful, quieter one disappears from his stockpile. She doesn't notice the microscopic distance opening up across the gravel because, as she goes forward – committing fully, going in hard, he takes a matching step back. But I see it all this time around and I feel for them both. I want to hold them, bash their stupid heads together, make her slow down and him speed up. Instead, one day I will teach her the back-to-back trick, to bridge the gap and hope to synchronise their heartbeats. But not yet. Though I know this story, I'm hazy on some of the finer details and I want to wait and give them a chance. I know why, of course, old romantic that I am. I want to see him spin and take her hands, smile broadly and look her in the eyes and meet her there, where she is and show her that it's okay to shrug some of this off. I long to see him walk around with his arms outstretched sometimes in case she falls – because she will – and then witness him scoop her up and kiss her as she sees how much he loves the shaky but real woman that I see. She doesn't need him to prop her up like this – I know that she is more than enough. Yet the me that was her is almost dissolved by wanting it to happen.

This is going to be hard, isn't it? I turn my eyes down to the gravel for a moment to let them clear. Weeds and lots of them, but it's probably best not to start focusing on that kind of thing as it's going to get a lot worse before it gets better. As I look at the grass between the stones and wish I could get my hori hori and start work, I notice another tool on the floor – the axe – and find I was expecting it. The settler's baby is here too, of course, and I heave him onto my back. There's a brass candlestick, a basket, a compass, a shoulder bag full of bulbs, the civil servant's report – a little grubby from where it's been lying – everything I am going to need, waiting for me: tools of a new trade. I am not finished with this place after all; I have another job to do. I imagine it will be lonely at first, taking comfort in the fiery treetops and talking to blackbirds until she is ready to open herself to the help I can give. And she is going to need a hell of a lot of help, isn't she? I shake my head and smile as she turns her naivety to face me and then, like an idiot, I wave. Almost before my hand moves I know it's stupid, but when you know someone as I know her and see them again after fifty-odd years, you can't help but get a little overexcited. She looks through me though, her gaze in a thousand other places, and I try to be okay with that. After all, she doesn't really see anything yet, just the blurry bigness of all that's ahead.

I move in a little closer – delighted to see the children's faces again, and to catch tiny glimpses of parts of his face too: the young man with strong arms and a big heart that he's keeping locked away for now. I love all three of these people, together they are the sun and I still orbit them in an ellipse of joy. And what about her: this bag of bones and notions, a ragtag collection of disguises stuffed into a body that she hasn't learned to find the edges of yet? Well, it seems I love her most of all. She is my life's work, my goal, my love story, my end result. The woman is my project, and I do like a project. And so does she, which is handy because, one day, I will be hers.

I see it spread out clearly now: how it all works, how it always worked. The us, the I, the me of it. The solid trunk I have wound myself back around, perhaps less mystical than I had presumed. It doesn't contain the secret of the universe, but then the universe probably has loads of secrets and why would it trust any of them to me? This thing I've been searching for, hunting down and crying over isn't a treasure made of interstellar dust and revelations. The treasure is and was always just myself – a little diffused and refracted perhaps – but me nonetheless. The shadow on the plot who guided me, voices of other times and places, the comfort, the knowledge I got from them and the things they shared: all me. The nudges showing me how to make things new for the now we needed, while not forgetting the lessons of before: also me. The prompt to put my hand on the clay and feel its atoms exert their force against my skin: me. In every one of those moments in my life, I thought I was too much or not enough and had to find the answers outside myself. I was wrong – it all came from the inside. Each lesson and every bit of hope delivered under this big sky with my fingers in the dirt was given out in the school of my own footsteps, my own maps. I sowed the seeds I saved. I dug, potted on and planted everything out myself and then, taking the blade in these very same hands, I hacked things back because they were over.

It makes me want to howl – with laughter, with anger and with frustration – as I turn to face her, the plot's new custodian, the thirty-four-year-old me, who will take more than a lifetime to discover this – that she's the answer to her own question. I am going to want to tell her but instead she will have to be shown. I look hard into her uncertain eyes. This is going to be quite the ride, kiddo, but we'll do it together. She doesn't see me yet, but I will keep trying until our eyes meet. Then she will notice it, what I have brought through the thread-end of death and out the other side to life. To a future I remember because I have already lived it, and a past that was mine, that I now hand on to her. All there ever was and all there will be: me – her – and the earth under our fingernails.

I reach out to take her pristine hand, keen, as ever, to get started, and then my eyes focus on the button-clicker that I'm holding in my weathered grip.

Red, Blue, Circle, Square –

Unfinished business.

I knew I hadn't completed that test. Yes, they took off the equipment, printed the results and told me it was over, but it's been running in the background all along, waiting for one final click. And now the moment has come. Fine. I think I remember how: press the button with my poised thumb, but only when I see the matching sequence. I wait for the test to re-start. This is my final scene and, when it is done, I will step back and cloak myself in shadows and stories.

The shapes start to slide across the screen. Red square, blue circle, red circle, blue circle, red square, blue circle, blue square, red, red, blue, blue, rose, delphinium, dahlia, anemone, bluebell, tulip, forget-me-not.

But this time I don't forget.

Red circle, blue circle, red circle, blue square. The shapes blend as they did decades ago but I make no attempt to keep up this time. I let them merge and muddle, seeing the primary-school truth that red mixed with blue makes purple. Red, blue, red, blue, red, rose, delphinium, foxglove, aster, iris, lavender, camassia.

Red circle, blue circle, red circle now,
finally,
the clarity of a purple circle.

My thumb moves to these flowers in the meadow –
mutaxsdi –
and then hovers as dark clouds fill the sky.

Thumb down:
press the button,
hold the lightening rod –
land firmly, Rebecca!

I feel the charge meet the earth one last time.

CLICK.

TEST END.

# AUTHOR'S NOTE

All the people who appear in this book are real and true to me, but as 'real' and 'true' mean different things, at different times, to different people I want to explain what these words mean to me.

I have tried to capture the story of the past eighteen months and snippets from the rest of my life as accurately and fairly as possible. But I know that everyone involved will have their own equally valid versions – after all, I have at least six versions myself and will likely have another six tomorrow. These pages are not intended to be definitive or give the last word on anything. Some names, places, place names and identifying details have been changed to protect my family and other individuals.

The historical figures, dates and references included here are all based on research and correspond to evidence I have found of individuals, places, times, events, traditions and movements. All the historical characters mentioned by name are documented to have lived by that name, at that time and in the place that I have written them as living in. The additional life I have breathed into these stories and women of the past is based on a mixture of research into what was likely or plausible and an instinct about what felt right. However, I am not a historian, a horticulturalist, a social anthropologist or a physicist – I might not have it all quite as it was, is or should be.

# ACKNOWLEDGEMENTS

Writing this book – about the worst times, during some of the worst times, in lockdown, with my kids off school and as I was being diagnosed and starting treatment – could very well have been the last straw. One the fundamental reasons it didn't finish me off (and was the source of joy and hope at times) is the belief, love, kindness, care, skill and dedication that has been shown to me by a publishing team of exceptional women. *Earthed* has made it this far because so many people have given it and me much more than they had to, at a time when all of us have had so little to give.

Many agents would have baulked at the idea of a client switching subject and genre so dramatically. Not Julia Silk. Her support, openness and relentless faith in me – through hard times and lengthy slogs – has transformed some of the worst years of my life into something that is starting to feel a little bit magic. Thank you, Julia. You are an exceptional agent and a fabulous woman and have cared for me, bolstered me, pushed me, told me things I needed to hear and always had my back. It is because of you that I have remembered I always wanted to be a writer and looked up to discover that I am one.

If I typed continuously for a week I wouldn't have written enough to adequately thank my editor Sarah Rigby. Sarah: you bought this book when it was something really quite different, seeing something in it that still lived in my subconscious and loving it from that first seed of an idea. Thank you for letting me run, spin and tumble with it and always being there, somehow knowing what I needed and giving me permission to experiment. Thank you for all the hours and all the tears – those mind-opening, soul-expanding conversations. The insight and skill that gave me nudged and sent my mind scattering in the right directions. I am incredibly grateful to you for mustering everything at your disposal (and more) to ensure I have been well-supported and the book given its chance. You have looked after me and held me safe in this process in every way you could. What a rare and precious thing that is.

I'd like to thank everyone at Elliott & Thompson, as well as the Simon & Schuster sales team, for their dedication and skill. Being published by you has been a wonderful revelation. Thanks to Pippa Crane for lightning speed and design wizardry; Marianne Thorndahl for supreme organisation and attention to detail; Emma Finnigan for enthusiasm and dedication to publicising my story, as well as being considerate and thoughtful as to its personal nature. I am grateful also to Lorne Forsyth for such a hands-on approach to caring for the authors you publish, as well as excellent advice on getting hens out of a laying strike!

I'm hugely inspired by Anna Fewster, whose cover design is a thing of great beauty and meaning. Anna: I've loved and admired you since that first message you sent me when we were both walking the Kent coastline with babies who wouldn't sleep on our backs. Thank you for working so incredibly hard and doing something so unique that wraps this book in something that has come from within. Your support team of Robert Stillman and Paul Fewster deserve a mention too. As does Penny

Fewster, for her gorgeous photography of the plot and thoughtful approach to a family and animal photoshoot.

My gratitude also to Nick Morley, Anna Fidalgo and Robin Crawford, who have all been roped in to helping at various stages of the editing and cover design process. I'm also grateful to Gary Sampson and everyone at the Woodchurch Ancestry Society for their painstaking work cataloguing parish records and researching the history of our village and its inhabitants. In particular, I would like to thank Josie Mackie for taking the time to share her knowledge, experience and memories with me and for leading me to Joane Newman. A huge thanks additionally to Bernard and Jean Smith for their generosity sharing photographs and memories of the plot's recent history.

I have only been able to get through these years because of the help of many medical professionals as well as the support of my friends, family and other writers. I'd like to express my gratitude to Jack Hickey and everyone at Woodchurch Surgery for their exceptional medical care and kindness over the past few years. To Max Zoettl, Julia and Ben Cherry who have all helped me hugely, as well as making time to read this book despite their busy professional lives. Thanks also to Jill Rowley for making my neck and back work again when eighteen-hour days writing about a breakdown left me incredibly tense and in pain.

Many friends have supported me since we moved to the plot. Clare Connerton: our friendship has been a marvellous gift and your support of my writing through Harbour Books is hugely meaningful. Robyn Wilder: you have pointed me and many others in the direction of a life-changing diagnosis, all the while suffering yourself. I am so glad to be near you even if we never manage to see each other. Thank you Laura Rees, Kieran Coyle, Leesa and Rich Holden, Ben Walgate, Jo Furneaux and Amy Jackson for the innumerable ways you've brightened our lives since we moved here. I am grateful to further-flung friends old and new for their love and care, especially: Eva Wiseman, Mark Ogus, Eloise Moody, Mars Lord, Emma Svanberg, Sophie Heawood, Catherine Dawson, Charlotte Philby, Maisie Hill, John Roynon and Dan Pacek. And to the friends and supportive professionals I have made through the animals and plants of our plot, including: Pam Bailey, Duncan Stewart, Jilly Raggett and Ylva Blid-Mackenzie.

When I started Mothers Who Write, I didn't realise how much the support and wisdom of other writers would benefit me while working on *Earthed*. I am particularly thankful to: Clover Stroud, Sarah Langford, Pragya Agarwal, Jess Moxham, Grace Timothy, Elizabeth Prochaska, Rachel Holmes, Susannah Okret, Sophie Fletcher, Amy Liptrot, Cariad Lloyd, Tara King and Marchelle Farrell. And to those who gave their permission for me to use the poems by Mary Oliver and Tania Hershman in this book.

With respect and thought, I have mentioned or written in more detail about some contemporary places, people and communities. I would like to acknowledge my debt to the G̲aandlee Guu Jaalang – the Daughters of the Rivers of the Haida first-nation people and their spokesperson, Kuun Jaadas (Adeana Young); Cheryl Bryce, knowledge keeper of the Lekwungen Songhees First Nation people; the Lokono people and Saamaccan people of today as well as their ancestors; the Tsawataineuk First Nation people who harvested muṯaxsdi in the past and also their descendants; the Bakongo and Kalinago peoples and the descendants of the Aztecs, the Nahual people; the people of the Ngarrindjeri and the people of the Kulin.

It is important also for me to acknowledge the tragedy of Ismail Mohamed Abdulwahab's death and express condolences to his family who were not allowed to be with their son in his last hours. And I want to record and protest the murder of George Floyd as well as to pay tribute to Alicia Garza, Patrisse Cullors and Opal Tometi – the founders of the Black Lives Matter movement.

I would like to thank the Landworkers Alliance, La Via Campesina, Land in our Names, Leah Penniman, Sara Limback, Justin Robinson, Remi Sade, Cel Robertson, Sui Searle, Claire Ratinon and all the organisations, grassroots groups and individuals I continue to learn from. I am grateful for all you do to advocate for an interlinked future of land, food, racial, climate and social justice.

Finally my family. Thanks and love to my parents for supporting this book so wholeheartedly even though reading it has been difficult. Thank you for giving me the dedicated time I needed to finish the first draft by looking after the children for three whole weeks and for the many ways you are trying to make life easier. The four of us are very lucky to have you both in our lives.

I am thankful to my brother for giving me permission to write about difficult, personal things and for embracing it. Emerging from this with a rebuilt relationship with you is something I am proud of and grateful for. Your practical help on the smallholding and your friendship are invaluable. It is exciting to see your life taking off.

Sofya and Arthur. I love you two more than I can ever say. Thanks for understanding and (mostly) respecting how much time I've had to spend on this book of late and for all the reasons you give me to be joyful and look at the world anew.

Jared: there aren't many words left are there? I used them all up on these pages past. Apart from thank you, perhaps. And I love you. I think we can do this if we continue to hold on to this place and to each other. Let's!